# THE STUDENT CHEMIST
EXPLORES ORGANIC COMPOUNDS

# THE STUDENT CHEMIST EXPLORES ORGANIC COMPOUNDS

By
*Ted Charney*

Illustrated by
*Nancy Lou Gahan*

RICHARDS ROSEN PRESS, INC.
New York, New York 10010

Published in 1976 by Richards Rosen Press, Inc.
29 East 21st Street, New York, N.Y. 10010

*Copyright 1976 by Ted Charney*

All rights reserved. No part of this book may be reproduced
in any form without written permission from
the publisher, except by a reviewer.

*First Edition*

*Manufactured in the United States of America*


Library of Congress Cataloging in Publication Data

Charney, Ted.
 The student chemist explores organic compounds.

 (Student scientist series)
 Bibliography: p.
 1. Chemistry, Organic. I. Title.
QD251.2C48     547     75-12904
ISBN 0-8239-0326-5

# *About the Author*

Ted Charney is Assistant Principal in charge of the Physical Science Department of William Cullen Bryant High School in Queens, New York. Before his present appointment, he taught chemistry at several high schools in New York City.

Mr. Charney's experience extends beyond the science classroom. He served for two years with the United States Army Chemical Corps. For six consecutive summers, he engaged in research work for the Lever Brothers Research and Development Center. As an educator, he has served on committees for writing citywide and statewide questions for the New York City Board of Education and the New York State Department of Education, and he has authored a series of slide sets for the teaching of atomic structures in high school chemistry courses. For the past four summers, Mr. Charney has directed the chemistry laboratory at a large children's camp in the Adirondack Mountains.

Mr. Charney has been a member of several scientific and professional associations, including the American

Chemical Society, the Chemistry Teachers Club of New York, and the Council of Supervisory Associations.

Because he enjoys science as well as teaching, Ted Charney feels he has the best of all possible worlds: that of a chemistry teacher and supervisor. He relishes the opportunity to lead students to the discovery that chemistry is a vibrant science, one that is both intellectually stimulating and intimately related to everyday experiences.

Mr. Charney was educated at the City College of New York and Adelphi University and holds B.S., M.A., and M.S. degrees. Under grants from the National Science Foundation, he also studied at Bowdoin College and Brooklyn Polytechnic Institute and took part in N.S.F. research projects at City College, specializing in organic synthesis.

Mr. Charney is married and has three daughters.

# Contents

**About the Author** — v

**Introduction**

**I. Organic Chemistry: What It's All About** — 3
*The Silent Majority — What's in a Name? — Vital Forces Lose Vitality*

**II. What Makes the Carbon Atom So Special?** — 7
*The Structure and Bonding of the Carbon Atom — Structural Formulas — Hybrid Orbitals: They Are All Mixed Up — The Valence Bond Theory and Orbital Hybrids*

**III. Diamond and Graphite: The Carbon Twins** — 19
*There Is More Than One Carbon — A Girl's Best Friend — An Old Softy — Untamed Electrons — Artificial Diamonds*

**IV. Hydrocarbons: The Happy Union of Two Elements** — 27
*What Kind of Fuel Am I? — Classes of Hydrocarbons — Saturated Hydrocarbons — Multiple Bonds — When You're Having More Than One — Electrons on the Run: Aromatic Hydrocarbons — Siamese Twins — With This Ring I Do Thee Wed — What It Means To Be a Hydrocarbon*

**V. Organic Relatives Of $H_2O$, $H_2S$, and $NH_3$** — 45
*One for the Road — King of the Smells — Organic Bases*

**VI. More Organic Compounds — For Life or Death** — 57
*From Carboxyl to Carbonyl — Bring on the Halogens — The Longest War: Man vs. Insect — The Sexy Insecticides — An Antibiotic That Kills Insects — Psychedelic Chemicals and Weed-Killers — They Are Not All Bad*

VII. *Isomers: It's How You Put It All Together That Counts*     81
*What It Means to Be an Isomer—Alkanes: One for All, All for One—Alkenes: Double Your Pleasure with a Double Bond—Cycloalkanes: The Nonconformers—Isomers, Rings, and Mothballs—Looking-Glass Isomers*

VIII. *Polymers: The Land of the Giants*     95
*Alkenes That Can Add—Copolymers: Where Molecules Live Together—Building a Building Block—Polymers Made of Sugar—Future Shock*

*For Further Reading*

# THE STUDENT CHEMIST
EXPLORES ORGANIC COMPOUNDS

# I

## *Organic Chemistry: What It's All About*

*The Silent Majority*

What is so special about the compounds of carbon that they should be studied separately from compounds of all the other elements of the Periodic Table? The answer is partly that the vast majority of the known compounds in our world contain carbon. There happen to be more compounds of this one element than of all the other 100 or more elements put together. The answer, too, is partly that carbon compounds are everywhere and touch our lives every single day. They can be found in our foods and clothing, our plastics and paints, our detergents and dyes, our medicines and fuels, and in our paper and ink. In addition, they also form the living matter of which our skin, muscles, blood, nerves, and tissues are composed. Carbon compounds are special also because they can be enormously large and complex in comparison to other compounds. Some molecules of carbon compounds contain thousands of atoms and can often react in ways that are far different from other compounds. A great deal of research in the area of organic chemistry is devoted to the study of these reactions.

Carbon compounds often play a dual role in our lives; they can do us much good as well as much harm. Modern insecti-

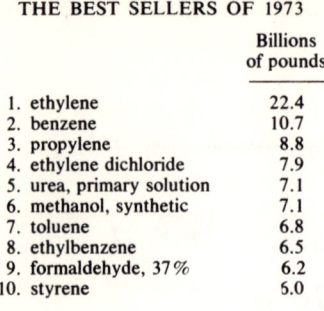

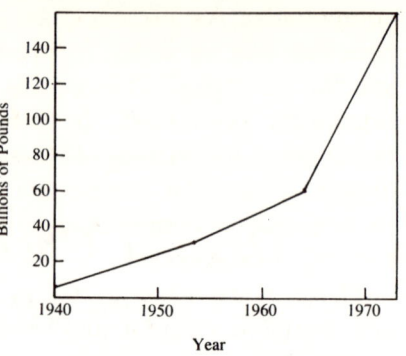

THE BEST SELLERS OF 1973

|   | Billions of pounds |
|---|---|
| 1. ethylene | 22.4 |
| 2. benzene | 10.7 |
| 3. propylene | 8.8 |
| 4. ethylene dichloride | 7.9 |
| 5. urea, primary solution | 7.1 |
| 6. methanol, synthetic | 7.1 |
| 7. toluene | 6.8 |
| 8. ethylbenzene | 6.5 |
| 9. formaldehyde, 37% | 6.2 |
| 10. styrene | 5.0 |

FIG. 1(a) *The ten most widely used organic chemicals in the U.S. in 1973.*

FIG. 1(b) *The annual production of synthetic organic chemicals.*

cides, for example, have prevented voracious bugs from taking over our planet (at least, so far), but they now are posing health problems so serious that some of our most effective insecticides have been banned. Powerful explosives, such as trinitrotoluene (TNT), nitroglycerin, and cyclonite, may be used as constructive instruments for peace or as destructive instruments of war. Psychochemicals, such as amphetamines, can relieve serious depressions of the mind or lead to the tragic consequences of addiction. Good or bad, the approximately 2,000,000 known carbon compounds are the reason we have organic chemists. The study of these compounds — their structures, their properties, their reactions, and their uses — is what organic chemistry is all about.

## What's in a Name?

The name organic chemistry was first applied to the study of the compounds that were obtained from plant or animal organisms, either alive or dead. At one time, these were the only sources of carbon compounds. It was on this basis, then, that organic chemistry was first associated with carbon

## Organic Chemistry: What It's All About 5

compounds. As far back as the time of Aristotle, it was believed that in order to produce an organic compound, a special or "vital" force was needed, which only living organisms possessed. It followed, then, that only living organisms, by virtue of their vital force, could produce organic or carbon compounds. Accordingly, only *inorganic* or non-carbon compounds could possibly be obtained from mineral sources.

There were several distinctions made between organic and inorganic compounds. Most of the earlier known organic compounds, such as methane and ethyl alcohol, were found to have relatively low melting and boiling points, poor electrical conductivity, and high flammability. On the other hand, such inorganic compounds as sodium chloride and silicon dioxide (sand), were high-melting-point solids that were incapable of burning. When melted or dissolved in water, inorganic solids are generally capable of conducting electricity. In chemical laboratories today, organic liquids, such as ether, methyl alcohol, and acetone, require extra precautions because of their volatility and flammability. Many gas stations have signs reminding drivers to shut off their automobile engines when purchasing gasoline, which is a complex mixture of organic liquids.

The concept of ion conductivity did not reach prominence until the late nineteenth century. Today, although many organic compounds are still isolated from plant and animal sources, most of them are synthetic or man-made. Synthetic compounds like nylon or DDT never encountered a living thing until they emerged from the flask, test tube, or vat. Organic compounds now are produced from both inorganic and organic sources. The terms "organic compound" and "carbon compound" now are used interchangeably regardless of the source. A small number of carbon compounds (cyanides, cyanates, carbonates, and bicarbonates) are still regarded as inorganic because of their mineral origin and because, unlike most organic compounds, they contain ionic bonds.

*Vital Forces Lose Vitality*

The belief in the vital force doctrine began to crumble in 1828, when the German chemist Wöhler synthesized the organic compound urea, a component of urine, from the "inorganic" ammonium cyanate:

$$NH_4NCO \xrightarrow{\text{heat}} CO(NH_2)_2$$

ammonium cyanate (inorganic)   urea (organic)

It was not until the 1850's that even the most die-hard vitalists conceded defeat. From then on, the production of organic compounds from inorganic sources became routine, and no taboos or mystical forces could deter the organic chemist from opening the doors of the chemical world to a whole variety of new and exciting compounds. The word "organic," although still used, is a relic of the old days when chemists believed in vital forces.

# II

# *What Makes the Carbon Atom So Special?*

*The Structure and Bonding of the Carbon Atom*

Carbon is just one of more than 100 known elements. What is so special about carbon, then, that enables it to form such a vast number of compounds? Why is carbon singled out above all other elements for this unique distinction? To a large extent, the answer to this question concerns the electronic structure of the carbon atom. From your exploration of the Periodic Table you found that carbon is located in Group IVA, period 2, and that it has an atomic number of six. Therefore, the atom of carbon has a total of six orbital electrons, two of which are in the K shell or first energy level and the remaining four of which are in the L shell or second energy level. The "L" electrons are the valence electrons of carbon and are the ones which participate in bonding. Inasmuch as atoms become more stable when they achieve the electron configuration of a noble gas atom, such as neon or argon, the carbon atom requires the addition of four more electrons to reach a total of ten electrons, thereby giving it the same electron configuration as the atom of neon. In order to gain the four extra electrons, which give it a total of eight electrons in the valence shell, the carbon atom forms four

## 8 THE STUDENT CHEMIST EXPLORES ORGANIC COMPOUNDS

bonds with atoms of other elements and even with other carbon atoms.

You have probably learned from your previous explorations of chemistry that the element carbon is neither strongly metallic nor strongly non-metallic in nature. Carbon atoms have neither a strong tendency to lose electrons nor a strong tendency to gain them in chemical reactions. Carbon's electronegativity (2.0 on the Pauling scale) and its first ionization energy (11.26 electron volts) are higher than those of such metals as sodium, calcium, and aluminum, but lower than those of such nonmetals as nitrogen, oxygen, and chlorine. What all this boils down to is that, under these conditions, the carbon atom usually neither loses nor gains elec-

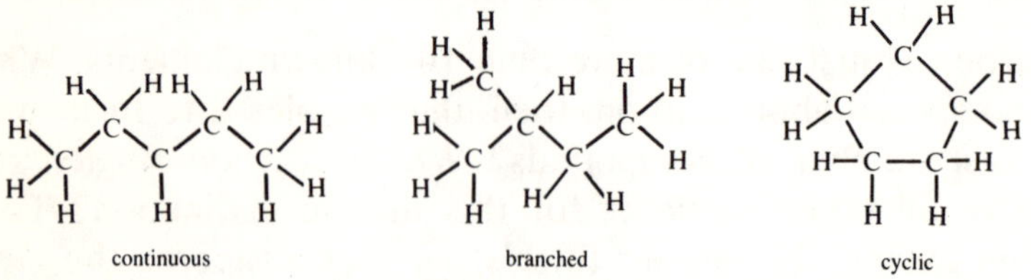

FIG. 2  *Types of molecular arrangements.*

trons in a chemical reaction with other atoms. Rather it shares electrons, which results in the formation of covalent bonds. In the overwhelming majority of compounds, carbon shares electrons with hydrogen atoms. Carbon atoms are also found to share electrons with atoms of oxygen, nitrogen, sulfur, phosphorus, and the halogens. There are some carbon compounds containing atoms of metals, such as tetraethyl lead. Silicones also contain the element silicon. In nearly all organic compounds, carbon atoms are also found to share electrons with other carbon atoms. This is one of its most unique properties. These other carbons, in turn, may bond with still other carbons in a great variety of ways. As shown in Figure 2, carbon atoms may link together in continuous

## What Makes the Carbon Atom So Special?

chains of varying length, in branched chains, or in cyclic (ring) arrangements.

A carbon atom can also form multiple bonds with other carbon bonds or with atoms of other elements, such as oxygen and nitrogen. In such bonds, the atoms may share four electrons (i.e., a double bond) or six electrons (a triple bond) between them. Examples of organic molecules containing multiple bonds are illustrated in Figure 3. A check of the number of electrons surrounding the carbon atoms in each example should reveal that a stable electron configuration with eight valence electrons has been achieved.

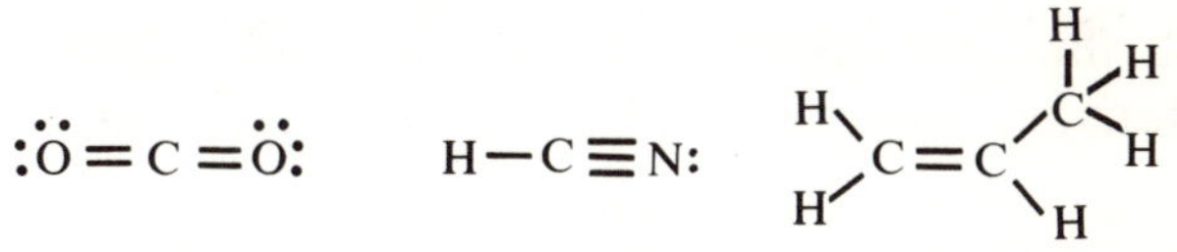

$$:\ddot{O}=C=\ddot{O}: \qquad H-C\equiv N: \qquad \begin{array}{c} H \\ \diagdown \\ H \end{array} C = C \begin{array}{c} H \\ | \\ C \\ \diagdown \\ H \end{array} \begin{array}{c} H \\ H \\ H \end{array}$$

carbon dioxide   hydrogen cyanide   propene

FIG. 3

There are over 3,000 compounds consisting of carbon and hydrogen alone. These compounds, called hydrocarbons, will be considered in more detail in Chapter IV. Boron, second only to carbon among period 2 elements in this respect, forms approximately fifteen boron-hydrogen compounds. When atoms of other elements are included in the organic molecule, the number of possible organic compounds rises sharply. The organic chemist devotes much time and effort to determining structures of compounds and finding ways to rearrange these structures in order to create new compounds. He is a molecular architect, so to speak, and can often custom-make molecules of a given size, shape, and property to meet a particular industrial, commercial, or research need. Figure 4, shows several strange and exotic-looking molecules that have been synthesized by the organic chemist. Unscientific names, such as cubane and prismane,

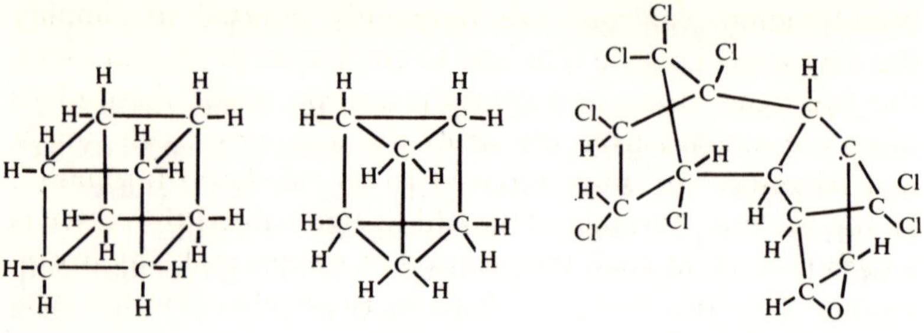

cubane    prismane    endrin
(a highly toxic chemical)

FIG. 4

are commonly used in lieu of their more scientific names because they are easier to remember and suggest certain characteristics about the molecules they represent.

## Structural Formulas

The diagrams shown so far are usually known as structural or graphic formulas. They supply visual information about the orientation of the atoms, the number and types of bonds and atoms, and the overall shape or geometry of the molecule. Molecular formulas, such as $C_8H_{18}$ and $C_6H_{12}O_6$, reveal only the chemical composition of the molecule. As we will see later, a given molecular formula may represent more than one specific organic compound. For example, there are eighteen different possible structures which may be represented by $C_8H_{18}$. Each dash in the structural formula represents a pair of shared electrons, that is, a covalent bond.

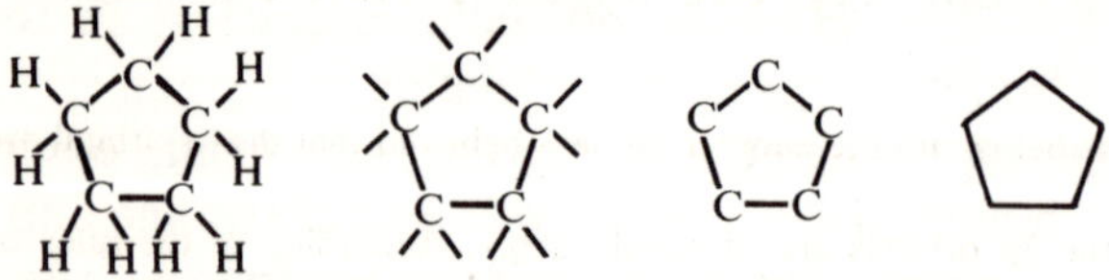

FIG. 5 *Different representations of cyclopentane,* $C_5H_{10}$.

## What Makes the Carbon Atom So Special? 11

Non-bonding electrons are frequently omitted to simplify the diagrams. To save time and to emphasize particular ideas, the hydrogen atoms and even the carbon atoms themselves often are omitted from the structural diagrams. Such simplified formulas are often referred to as "skeleton formulas." In the skeleton formula, it should be understood that there is a carbon atom at each intersection of dashes and a hydrogen atom at the open end of each dash. Remember that the total number of hydrogen atoms, written or unwritten, must satisfy the valence requirements of the carbon atoms.

### *Hybrid Orbitals: They Are All Mixed Up*

Let us now consider in more depth the molecule of methane, $CH_4$. This is the simplest of all organic molecules. In particular, let us consider the distribution of carbon's six electrons among the three atomic orbitals in the ground state configuration of the carbon atom in methane. Atomic orbitals, you may recall, are regions of space in which an electron is most likely to be found. Starting from the orbital nearest the nucleus and proceeding outward, the atomic orbitals are designated 1s, 2s, 2p, 3s, 3p, 3d, and so on.[1] This sequence also represents the orbitals in order of increasing energy. The configuration in Figure 6 shows that the carbon atom has an unpaired electron in two of the three 2p orbitals.[2] On this basis, the carbon atom might be expected to bond with only two hydrogen atoms, since two electrons are already paired off in the valence shell and are, therefore, unavailable for bonding with the lone electron on each hydrogen atom. The two unpaired electrons would pair off

---

[1] In atoms below atomic number 21, it is believed that the 4s orbital precedes the 3d orbital.

[2] The three 2p orbitals are of equal energy. According to the rules of quantum mechanics, an orbital may not be occupied by a pair of electrons until all other orbitals of equal energy are each occupied by one electron.

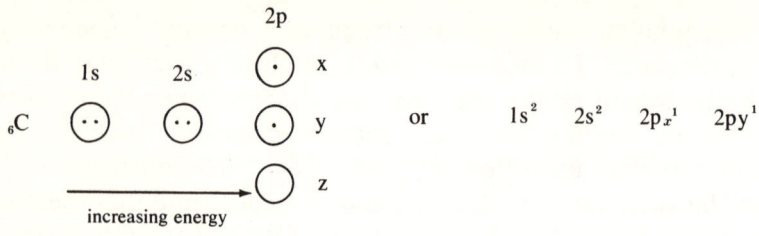

FIG. 6 *Orbital configuration of carbon atom in the ground state. The dots represent the electrons.*

with unpaired hydrogen atom electrons, giving rise to a formula of $CH_2$ for methane. Why then is the formula $CH_4$? Apparently, we must reexamine our model of the carbon atom to account for the four *equivalent* bonds which carbon has in the methane molecule. Equivalent bonds are bonds of equal energy, and, as we have noted in Figure 6, the four valence electrons of the ground state carbon atom are not all equal in energy and would not form equivalent bonds. How do we explain these inconsistencies?

## The Valence Bond Theory and Orbital Hybrids

In 1935, Linus Pauling, subsequently the winner of two Nobel Prizes, introduced the *valence bond theory* to account for the geometry of covalently bonded molecules. According to this theory, a molecule is made up of *atoms* whose valence shells overlap, resulting in a sharing of electrons. In other words, the atoms in the molecule act like isolated atoms, except that the electrons in the valence shell of one atom are also accommodated in the valence shells of adjacent atoms. In order to accomplish this sharing of electrons, the atomic orbitals of the valence shells of these atoms must *overlap* (Figure 7). Another theory, the *molecular orbital theory*, considers the molecule to be made up of *nuclei* of

# What Makes the Carbon Atom So Special?

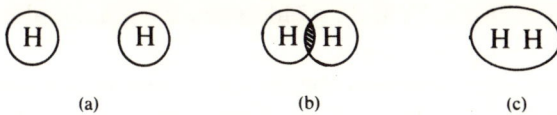

FIG. 7 Overlapping of the 1s orbitals of two hydrogen atoms to form a molecule of $H_2$.

atoms, each atom contributing all its electrons to a common pool. These electrons move about throughout the entire molecule in molecular orbitals rather than in individual atomic orbitals and belong to the entire molecule rather than to individual atoms. Because the valence bond approach is somewhat simpler than the molecular orbital approach, it remains in great use. By applying the concept of *orbital hybridization,* the valence bond theory successfully accounts for the fact that a carbon atom forms four equivalent bonds in methane (and in many other molecules as well).

According to this concept, the carbon atom must become "excited" by the promotion of some of its electrons to higher energy states or levels. Suppose we first promote one of the two 2s electrons to the vacant 2p orbital. This gives us the configuration shown below, which reveals a total of four unpaired electrons. We are now one step closer to the actual model of methane because now we would need four hydrogen atoms to complete the bonding requirements instead of only two. However, this would still not give us four equivalent

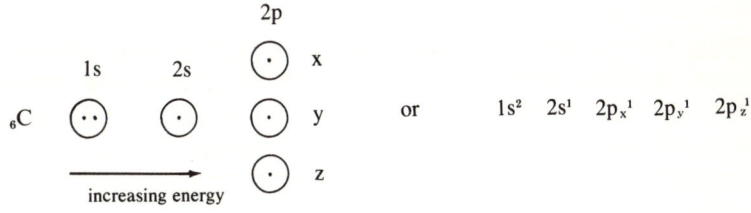

FIG. 8 Orbital configuration of carbon atom with one electron "promoted" from the 2s orbital to the vacant 2p orbital.

bonds because the 2s and 2p electrons are not equal in energy. So another step is needed.

The next step involves the *hybridization* of the one 2s and the three 2p orbitals so that all orbitals in the valence shell are made equal in energy. The term "hybridization" implies a mixing or blending process, just as in genetics where the gene for one trait and the gene for another trait mix to form a hybrid organism. In the case of the carbon atom, the 2s and 2p orbitals are all blended together to form four new *hybrid orbitals* that are equivalent. These hybrid orbitals are called $sp^3$ orbitals because they were formed by the mixing of one s and three p orbitals. The superscripts indicate the number of parent orbitals used to form the hybrid. The four $sp^3$ orbitals are centered about the nucleus of the carbon atom with their large lobes directed toward the corners (apices) of a regular tetrahedron. Figure 9 illustrates the hydridization process, and Figure 10 shows the shape and tetrahedral orientation of the four $sp^3$ hybrid orbitals.

By directing the orbitals to the corners of a tetrahedron and assuming that only one valence electron of carbon will most likely occupy each of the four orbitals, a minimum of electrical repulsion and greater stability are achieved. With this model of the carbon atom in mind, let's proceed to the last step in the formation of the methane molecule.

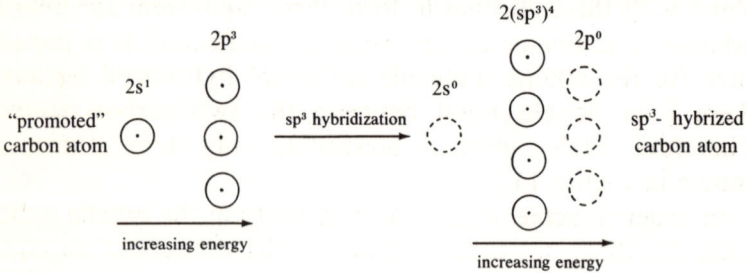

FIG. 9 *Orbital configurations of a "promoted" carbon atom undergoing $Sp^3$ hybridization.*

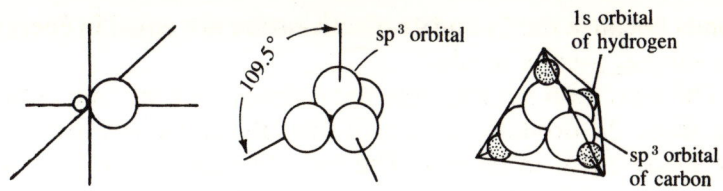

Fig. 10 (a) *shape of an* sp³ *hybrid orbital*
(b) *and* (c) *orientation of* sp³ *orbitals in methane*

Four hydrogen atoms unite with the carbon atom by overlapping their 1s orbitals with the four hybrid orbitals of the carbon. Now we have the four equivalent carbon-hydrogen bonds present in the methane molecule and have successfully accounted for methane's tetrahedral shape.

There are many other molecules in which the carbon atoms are sp³ hybridized as in the case of methane. Carbon tetrachloride, $CCl_4$, for example, is very similar in shape and structure to methane. Also, in molecules such as ethane ($C_2H_6$) the bonds to each carbon are sp³ hybridized and oriented tetrahedrally about each carbon nucleus.

There are many compounds, however, in which not all the p orbitals are involved in hybridization. For example, the carbon atoms in the ethene molecule ($C_2H_4$) are sp² hybridized. This means that only two of the three 2p orbitals blend with the 2s orbital to form three equivalent sp² hybrid orbitals. The remaining 2p orbital is unaffected. It is impossible for the ethene molecule to be sp³ hybridized because there is a double bond between the two carbon atoms. Therefore, each carbon is bonded to only three atoms, as shown in Figure 11.

In order to achieve maximum stability in the ethene molecule, the three sp² orbitals point to the corners of an equilateral triangle. The resulting bond angles about the carbon atoms are 120° as compared to approximately 109° in the

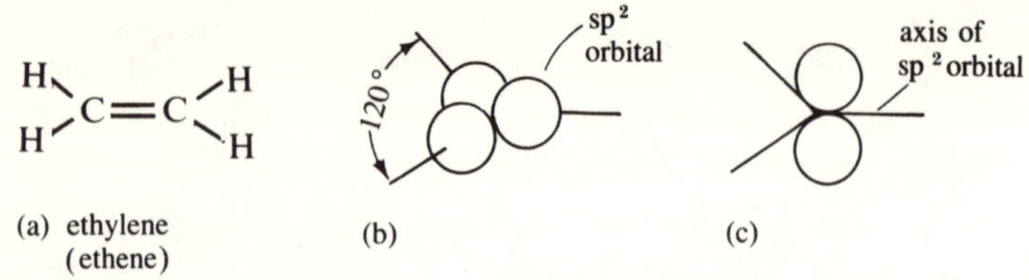

16 THE STUDENT CHEMIST EXPLORES ORGANIC COMPOUNDS

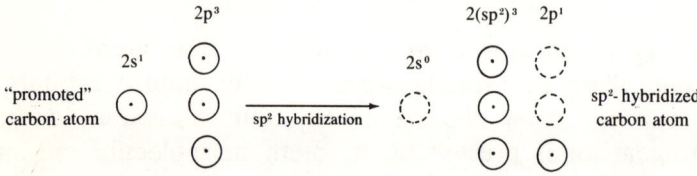

(a) ethylene (ethene)    (b)    (c)

FIG. 11 *Orientation of orbitals in* $sp^2$ *hybridized carbon atom.*
(b) $sp^2$ *orbitals only;* p *orbitals omitted*
(c) $sp^2$ *orbitals omitted;* p *orbitals are perpendicular to axes of* $sp^2$ *orbitals*

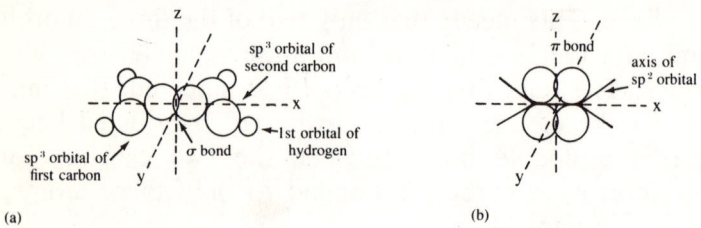

FIG. 12 *Orbital configurations of a "promoted" carbon atom undergoing* $sp^2$ *hybridization.*

tetrahedrally oriented $sp^3$ orbitals in methane. The $sp^2$ hybridization process is pictured in Figure 12.

The unhybridized 2p orbital of carbon is directed perpendicularly to the plane of the triangle with one of the dumbbell-shaped lobes extending above the plane and the other extending below, as shown in Figure 13. The double

FIG. 13 *Orientation of orbitals in a molecule of ethylene (ethene).*
(a) *showing only the* $sp^2$ *orbitals of carbon overlapping with the* 1s *orbitals of hydrogen; the unhybridized* p *orbitals are not shown*
(b) *the unhybridized* p *orbitals, one from each carbon, overlap to form a pi* ($\pi$) *bond; the* $sp^2$ *orbitals are not shown*

# What Makes the Carbon Atom So Special?

bond in ethene is formed in part when one of the sp² orbitals from one carbon atom overlaps end to end with one of the sp² orbitals from the other carbon. The end-to-end type of overlap is said to produce a *sigma* type of covalent bond. Simultaneously, the unhybridized p orbitals, one from each carbon, overlap side to side or edge to edge, forming what is called a *pi* bond. The sigma and pi bonds are named after Greek letters and are often represented by the Greek symbols σ and π, respectively. The double bond, therefore, consists of one sigma and one pi bond.

In the molecule of ethyne ($C_2H_2$), each carbon is *sp* hybridized; that is, only one 2s and one 2p orbital are blended together to form two hybrid orbitals. In ethyne, there is a triple bond between the two carbon atoms and each carbon atom is bonded to only one other atom, as shown in Figure 14. The sp orbitals are pointed in exactly opposite directions,

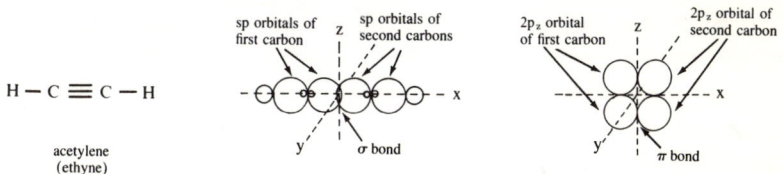

FIG. 14 *Orbital picture of acetylene (ethyne).*
(a) *unhybridized p orbitals are not shown*
(b) sp *orbitals are not shown*

making bond angles of 180°. The two remaining 2p orbitals are at right angles to each other and to the sp orbitals. In forming the triple bond, one of the sp orbitals and both 2p orbitals from each carbon atom overlap. The remaining sp orbital from each carbon overlaps with the 1s orbital of the hydrogen atom. The ethyne molecule, therefore, contains three sigma bonds (the two carbon to hydrogen bonds and one of the bonds in the triple bond) and two pi bonds (the other two bonds in the triple bond).

The concept of orbital hybridization can be applied to other elements besides carbon. When applied to atoms of oxygen and nitrogen, for example, the geometric shapes and bonding of the water and ammonia molecules can be better understood.

# III

## Diamond and Graphite: The Carbon Twins

### There Is More Than One Carbon

The element carbon itself exists in more than one form or structure. When an element appears in more than one form, the phenomenon known as *allotropy* occurs, and the different forms of the element are themselves known as *allotropes* or *allotropic forms*. Many other elements exhibit allotropy. Oxygen, for example, normally exists as ordinary molecular oxygen ($O_2$) and less commonly as ozone ($O_3$). Although the molecules of oxygen and ozone each contain a different number of oxygen atoms, the atoms themselves are identical. Consequently, the chemical behavior of the two allotropic forms of oxygen is very much alike. As proof of this, one can burn a piece of carbon in ordinary oxygen and a similar piece in ozone and, in each case, obtain the same product — carbon dioxide.

$$C + \underset{\substack{\text{ordinary} \\ \text{oxygen}}}{O_2} \rightarrow \underset{\substack{\text{carbon} \\ \text{dioxide}}}{CO_2} \qquad 3C + \underset{\text{ozone}}{2O_3} \rightarrow \underset{\substack{\text{carbon} \\ \text{dioxide}}}{3CO_2}$$

In fact, one allotropic form of oxygen may be converted into the other, as shown by the following equation:

$$3O_2 \rightleftarrows 2O_3$$

## 20 THE STUDENT CHEMIST EXPLORES ORGANIC COMPOUNDS

We now can explain the sudden increase in atmospheric ozone during a lightning storm. With the energy supplied by the lightning, some molecules of $O_2$ split apart and rearrange to form molecules of $O_3$.

Carbon also exists in two allotropic forms: diamond and graphite. It is somewhat hard to imagine that these two substances, which have such different properties, are actually composed of the same kind of atom. Pure diamond is colorless, very hard, and has little electrical conductivity. Graphite, on the other hand, is black, soft, and a good conductor of electricity. Diamond is extremely dense, having a density of 3.5 grams per milliliter (ml). The density of graphite is only 2.3 grams per ml. How, you may wonder, can the same element exist in such widely divergent forms? Inasmuch as both diamond and graphite consist of identical carbon atoms, their differences must stem from the way the carbon atoms are arranged in their crystalline structures.

### A Girl's Best Friend

In diamond, each carbon atom is surrounded by four other carbon atoms.[1] Each carbon becomes the center of a tetrahedron, with the adjacent four carbon atoms at the corners.[2] Each corner carbon atom is also the center of another tetrahedron. Continuing this process, with a great number of atoms we obtain the orderly arrangement of carbon atoms that characterizes the structure of a diamond crystal. Every diamond in the world, even one as large as your fist, is actually a single molecule because every atom in it is bonded either directly or indirectly to every other atom.

The extreme closeness of the atoms accounts for the high density of diamond, its hardness, and its ability to refract

---
[1] Only those carbon atoms at the surface of the diamond will not be bonded to other carbon atoms because not enough carbon atoms are available.
[2] The carbon atoms in diamond are sp³ hybridized.

## Diamond and Graphite: The Carbon Twins 21

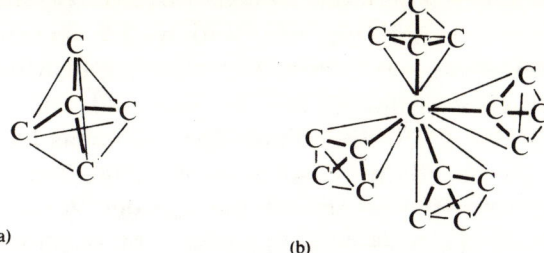

FIG. 15 *Orientation of carbon atoms in diamond.*
(a) *tetrahedral arrangement of carbon atoms about a single carbon atom*
(b) *groupings of tetrahedra in a section of a diamond crystal*

(bend) light very strongly. As the light penetrates the diamond, each colored component of the light is refracted to a different extent. This refraction and dispersion (separation) of light produce the glitter that makes the diamond so appealing as a gem. Emerging light rays striking a polished surface of a diamond at any angle greater than 24.5° undergo internal reflection, which adds to the brilliance of the gem.

## An Old Softy

Graphite is also a highly ordered, crystalline solid, but it has a different arrangement of atoms. In graphite, the carbon

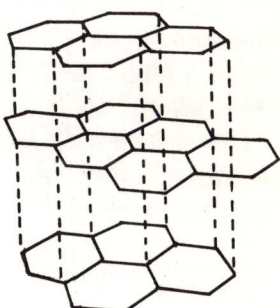

FIG. 16 *The layers of carbon atoms in a section of graphite.*

atoms are situated in sheets or layers which have little attraction for one another (Fig. 16). Within each layer, however, the carbon atoms are even closer together than in diamond and have very strong bonds. In diamond, the bonds are equally strong in all three dimensions, whereas the bonds in graphite are essentially two-dimensional. The weak attractive forces between the layers (of the van der Waals type) account for the softness and slipperiness of graphite. For this reason, graphite is used as a lubricant and in "lead" pencils.[1] When you make a stroke with a lead pencil on a sheet of paper, you apply some pressure which causes the layers of carbon atoms to slide apart and streak off.

## Untamed Electrons

Inspection of the structures of graphite and diamond reveals another major distinction. In diamond, each carbon atom is bonded to four other carbon atoms; in graphite, the number is only three. If each carbon atom in graphite shares one of its four valence electrons with each of its three neighboring atoms, there would still be an electron left unshared. This would give a total of seven electrons in the valence shell of each atom, a condition which we know is not stable. Yet, we know that graphite is more stable under ordinary conditions than diamond. Therefore, we can assume that the carbon atoms in graphite somehow manage to obtain the usual octet of electrons. But how? Explanations of natural phenomena are not always simple, and the chemist must often resort to his fertile imagination to come up with new and untraditional ideas.

Could each carbon atom form a double bond with one of its three neighbors? If so, which one? Because all three neighboring carbons are identical, it is unlikely that this would

---

[1] The ordinary lead pencil does not contain the element lead, unless it is in the paint covering the wood.

happen. Now turn up your power of imagination. Could each carbon atom donate its unshared fourth electron to a "pool" in which all the atoms in the entire layer can participate? These electrons would be *delocalized* and be able to roam about freely from atom to atom within the layer. Is there evidence to support this theory? Yes; a similar situation occurs in metallic crystals. As a result, graphite, too, is capable of conducting an electric current. Graphite even has a grayish sheen which makes it look somewhat like a metal and made it once be mistaken for the element lead.[1] The electrons in diamond, on the other hand, are all involved in bonding and are unavailable for conducting electricity. If you dissect a flashlight battery (dry cell), you will find that one of the electrodes is a black solid—graphite!

*Artificial Diamonds*

Both diamond and graphite, being allotropic forms of carbon, burn in an ample supply of oxygen to form carbon dioxide:

$$C_{(diamond)} + O_2 \rightarrow CO_2 \qquad C_{(graphite)} + O_2 \rightarrow CO_2$$

Furthermore, if diamond and graphite are different forms of the same element, should it not be possible to convert one into the other, as in the case of oxygen and ozone? If you have an old, unwanted diamond, you could melt it if you had a special oven that was able to reach a temperature of approximately 3500° C. When the material cooled off, you would find that the diamond had changed into graphite.

It is not likely that you would want to convert a precious diamond into inexpensive graphite. On the contrary, men had long dreamed and hoped that, someday, someone would change graphite into diamond. If nature can make a diamond,

---

[1] Hence the use of the name "lead" pencil.

why can't man? Nature produces diamonds more than 100 miles below the earth's crust, where the temperature and pressure are high. The diamonds mined today were probably carried up to the crustal region by eruptions of water or liquefied gas trapped deep in the earth when the diamonds were formed.

Diamonds have also been produced by the enormous impact of meteorites striking the earth. Scientists have long been able to produce sufficiently high temperatures to synthesize a diamond, but not the tremendous pressure needed. The breakthrough came in 1955, when research scientists at the General Electric Company triumphantly announced they had succeeded in making artificial diamonds every bit as good as nature's own, only smaller. They placed some powdered graphite and a pinch of metal catalyst into a container the size of a green pea and then concentrated the might of a huge steel press on this tiny target. The press squeezed the carbon atoms in graphite together until it reached a pressure of 1,800,000 pounds per square inch—a pressure equal to that exerted by 240 miles of earth. The graphite was heated to 4400° F at the same time to help break the bonds between the carbon atoms in the layers. After all this, among the residue of the gritty, black unchanged graphite, tiny diamonds shone. Man had duplicated one of nature's mightiest feats!

$$C_{(graphite)} \rightarrow C_{(diamond)}$$

On another occasion, scientists tried to make artificial diamonds by substituting high-carbon substances for graphite, which is pure carbon. They succeeded in this, too. Using their high-pressure machine, they actually obtained diamonds from peanut butter. Can you imagine purchasing diamonds that are either creamy or crunchy style?

Artificial diamonds are used in industry mostly for drilling and grinding operations because of their extreme hardness and durability. Diamond phonograph needles, for example, outlast other types of needles. Diamond-studded bits are used by dentists for drilling teeth and by oil companies for drilling new wells.

It was once thought that there were still other forms of carbon, collectively called the "amorphous" or non-crystalline forms. These forms, including coke, animal and wood charcoal, lampblack, and carbon black, have been found to be microcrystalline varieties of graphite.

As interesting and important as the element carbon is, either as diamond or graphite, the investigation of carbon's compounds is vastly more stimulating and far-reaching in application.

# IV

# *Hydrocarbons: The Happy Union of Two Elements*

*What Kind of Fuel Am I?*

Before attempting to understand the larger, more complex molecules, let us first explore a few things about the simplest organic compounds, the hydrocarbons. Although the hydrocarbons consist of only two elements, carbon and hydrogen, they occur in thousands of various shapes, sizes, and properties. Hydrocarbons are used more widely than any other class of organic compound. Each person in the United States uses more than 40 pounds of hydrocarbons in a single day!

The combustibility of hydrocarbons is chiefly responsible for their main application—as a fuel or source of energy. Hydrocarbons, in some form, power our automobiles, warm our homes in winter, and supply the energy for countless other needs. Hydrocarbons are chiefly obtained from the residues of animal and vegetable matter that were buried in the earth long ago. These residues, which include natural gas, petroleum, and coal, constitute the major sources of energy in most countries today. The dwindling supply of these sources of energy is largely responsible for our energy crisis. New energy sources will have to be explored. Some, which are being developed now and will be utilized to a greater extent in the future, are mechanical (moving air and water), chemical (fuel cells), solar, nuclear, and geothermal (earth heat) sources of energy. It should not be too surprising, therefore, if in the near future we find windmills as well as TV antennas on top of our homes, factories, and office buildings. Some scientists are experimenting with hydrogen as a substitute for the fossil fuels and even predict that hydrogen may replace these fuels as a primary energy source before the end of the century.

## Classes of Hydrocarbons

Hydrocarbons are classified according to the type of carbon-to-carbon bonds that the molecules contain: single, double, or triple. Those with only single bonds are said to be

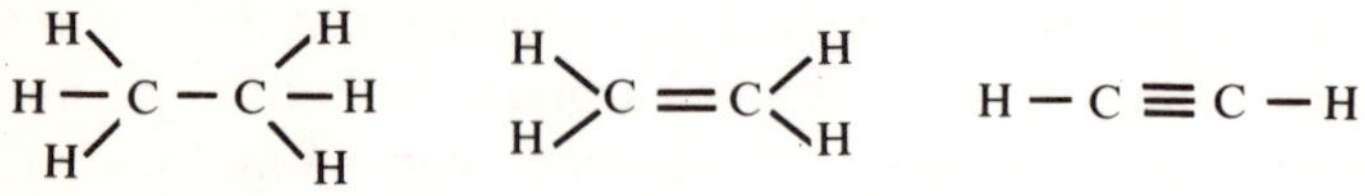

$$H_3C-CH_3 \quad\quad H_2C=CH_2 \quad\quad H-C\equiv C-H$$

ethane
a saturated
hydrocarbon

ethene

ethyne

unsaturated hydrocarbons

FIG. 17 *Classification of hydrocarbons according to type of bond.*

saturated because they contain the maximum number of hydrogen atoms possible. Those with double and/or triple bonds are described as unsaturated.

Another method of classifying hydrocarbons depends on the manner in which the carbon atoms are arranged. *Aliphatic hydrocarbons* are those which contain open chains (straight or branched) of carbon atoms. *Alicyclic hydrocarbons* contain rings of carbon atoms. These two types of hydrocarbons may be saturated or unsaturated. A third type of hydrocarbon, the *aromatic hydrocarbons,* also contains rings of carbon atoms, but these rings are never saturated and

# Hydrocarbons: The Happy Union of Two Elements

are structured so that electrons are delocalized and may wander about the ring much as they do in the graphite crystal. The delocalization of electrons in aromatic hydrocarbons tends to make these molecules more stable and, consequently, relatively inert to many of the chemicals with which the aliphatics and alicyclics react. This method of classification is also applied to other organic compounds in addition to the hydrocarbons.

## Saturated Hydrocarbons

Methane, $CH_4$, is the simplest possible organic compound and, therefore, is necessarily the simplest hydrocarbon. It is the chief ingredient of natural gas and the first member of

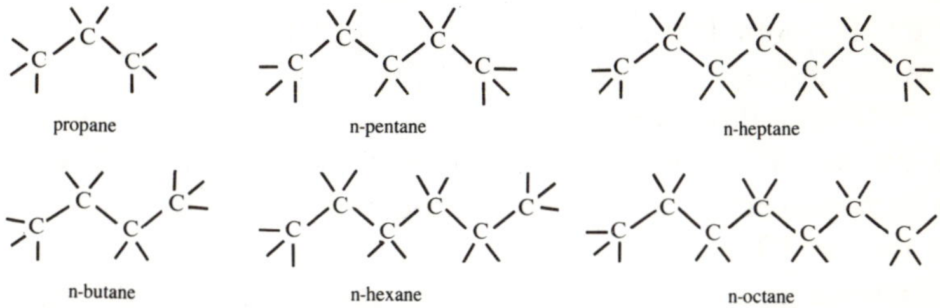

FIG. 18 *Members of the alkane series from $C_3$ to $C_8$. The "n" refers to the continuous chain nature of the molecules shown.*

a series of structurally related (homologous) compounds called *alkanes*. The alkanes are aliphatic hydrocarbons which may be represented by the general formula $C_nH_{2n+2}$, where "n" represents the number of carbon atoms and "2n + 2" the number of hydrogen atoms.

Some countries are now producing methane in sizable quantities from solid animal wastes. Factories in India produce methane from cow manure. Many families in Taiwan actually cook their meals with methane produced from hog manure. In fact, ten pigs can furnish enough methane to cook the food for a family of four. Curbing pollution in this manner may not be very aesthetic, but it is highly economical.

A mixture of approximately 90 percent propane and 10 percent butane is liquefied under pressure and sold in steel tanks as "liquefied petroleum gas" (LP gas). The shortage of gasoline and its sky-rocketing cost have spurred interest in LP gas as a possible alternative fuel. Engines powered by LP gas have been used in transit buses, delivery trucks, taxis, and hundreds of thousands of automobiles. Compared to gasoline as a fuel, LP gas burns cleaner, requires less frequent changes of spark plugs, is possibly safer, and currently is less expensive. On the debit side, LP gas delivers slightly fewer miles per gallon in most automobile engines. Only in engines capable of burning high octane fuel (octane rating of LP gas is about 110 to 115) efficiently can LP gas deliver the same gas mileage as gasoline. Whether or not the automobile you will someday own is powered by LP gas will depend on the supply of gasoline at the time and on the willingness of the auto makers to manufacture modified automobile engines that run on LP gas.

Gasoline is a mixture of alkanes ranging in composition from six to twelve carbon atoms per molecule. It is obtained by distilling petroleum, a far more complex mixture of organic compounds (mostly alkanes), and collecting the fraction that boils between 70° C and 200° C. The gasoline is then refined in order to remove any unsaturated hydrocarbons (which tend to form sticky substances called gums) and noxious sulfur compounds (which tend to reduce the efficient burning of the gasoline).

Various chemicals are added to gasoline in controlled amounts to improve its quality. Aromatic hydrocarbons raise the gasoline's octane rating, which indicates how well the gasoline performs in a test engine. Tetramethyl lead, $(CH_3)_4$ Pb, and tetraethyl lead, $(C_2H_5)_4$ Pb, are organometallic compounds. They act as catalysts to reduce the engine knocking caused by ill-timed ignition of the gasoline vapor. Dibromoethane, $C_2H_4Br_2$, is added in conjunction with the organolead compounds to prevent the formation of harmful lead deposits. Leaded gasolines are colored by the addition of various dyes to call attention to their highly toxic contents. There is an average of 2.5 ml of organolead compounds per gallon (3,785 ml) of leaded gasoline. About three fourths of the lead ultimately appears in the exhaust. Prolonged exposure to relatively small quantities of lead may cause serious illness or even death. Lead enters our bodies from food, dust, and drink as well as from air. The cigarette smoker gets an added bonus: each puff he inhales contains approximately 0.3 micrograms of lead.

Kerosene, like gasoline, is a mixture of alkanes which are in the $C_{12}$ to $C_{15}$ range. It is used extensively as jet-propulsion fuel. A typical jet plane today consumes about 300 gallons of fuel just to taxi from the warm-up ramp to the end of the takeoff runway. You have probably noticed the tremendous outpouring of pollutants each takeoff dumps into our skies. Diesel fuel, in the $C_{15}$ to $C_{18}$ range, is used in domestic heating and in the diesel engines of big trucks and boats. The bigger alkane molecules are capable of producing more energy per gallon of fuel oxidized. Diesel fuel is more economical and safer to use than gasoline. It, too, is a possible alternative to gasoline if there is a sufficient supply. At the present, however, much diesel fuel is "cracked" or decomposed to form smaller alkanes that fall into the gasoline range. Without such supplementary sources, the use of gasoline would have to be curtailed far more than it has been.

Larger members of the alkane series serve as lubricants or waxes. Still larger alkanes constitute the asphalt and tar employed for roads and roofing. Some alkanes, in the $C_{27}$ to $C_{35}$ range, occur naturally in plants and animals. In plants, they are most abundant in the waxes that act as protective coatings on leaves and stems and help to decrease evaporation. It is a curious fact of nature that most of these plant wax molecules have odd numbers of carbon atoms, usually 27, 31, or 33.

All alkanes are saturated hydrocarbons of the aliphatic variety. Given three or more carbon atoms per molecule, other saturated hydrocarbons can be formed if the carbon chain closes to form a ring. These "cyclic alkanes" are called *cycloalkanes*. The shape and structure of the three simplest cycloalkanes are shown in Figure 19. The shape and structure of cyclohexane will be discussed in Chapter VII.

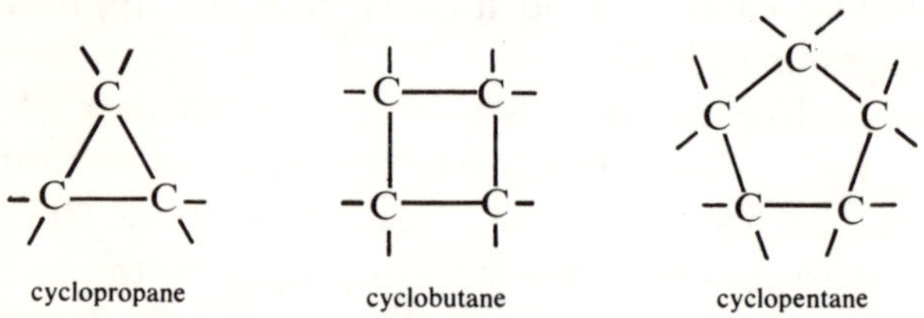

cyclopropane        cyclobutane        cyclopentane

FIG. 19   *Cycloalkanes from $C_3$ to $C_5$.*

Cyclopropane is a sweet-smelling, colorless gas which is used as an anesthetic in many types of surgery. With only three carbon atoms per molecule, it is the smallest possible cyclic compound. However, cyclic compounds may also be complex. Some contain as many as fifty atoms. A few, such as those found in certain viruses, consist of molecules in which hundreds of thousands of atoms of different elements are joined together in twisted strands, as in a necklace of microscopic beads. Cyclopentane and cyclohexane are of great industrial importance because they can be converted to aromatic hydrocarbons and hydrogen. The aromatic hydrocarbons, in turn, are used to produce high octane gasoline. Hydrogen serves in the purification of the gasoline and the synthesis of ammonia. There are also "polycyclic" hydrocarbons containing two or more rings with one or more carbon atoms in common. One such compound is decalin, $C_{10}H_{18}$, which has ten carbon atoms arranged in two cyclohexane rings joined together like Siamese twins.

## Multiple Bonds — When You're Having More than One

The cracking of alkanes also produces many unsaturated hydrocarbons. These are used as cheap raw materials for a large variety of important products, including plastics, solvents, lacquers, resins, and synthetic rubber. If, hopefully, alternate sources of energy can be found, the fossil fuels will be valued more as raw materials for organic syntheses than as fuels.

In unsaturated hydrocarbons, two carbon atoms share more than one pair of electrons. The *alkenes* are unsaturated hydrocarbons in which two adjacent carbon atoms share two pairs of electrons, thus forming a double bond. Ethene, $C_2H_4$, also called ethylene, is one of the most important alkenes. Over 13,000,000,000 pounds of it are produced each year. It is employed in the manufacture of polyethylene, a commonly used plastic, and ethylene glycol, the major component of most types of antifreeze for automobile radiators. Ethylene also has the unique facility of stimulating the ripening of fruits. In one recent experimental program, a six-membered cyclic compound called cycloheximide was applied to oranges still too tightly secured to the stem for harvesting. The addition of cycloheximide stimulated the oranges to produce ethylene, which, in turn, stimulated ripening and allowed the fruit to be detached easily from the stem.

Hydrocarbons containing carbon-to-carbon triple bonds are called alkynes. Ethyne or acetylene, $C_2H_2$, burns in oxyacetylene torches to produce the high temperature required for cutting and welding metals. Like the alkenes,

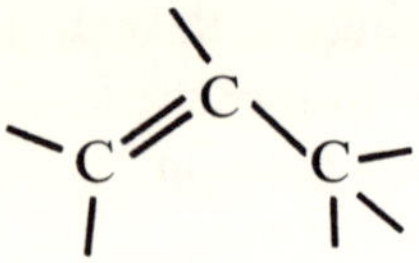

34 THE STUDENT CHEMIST EXPLORES ORGANIC COMPOUNDS

propene

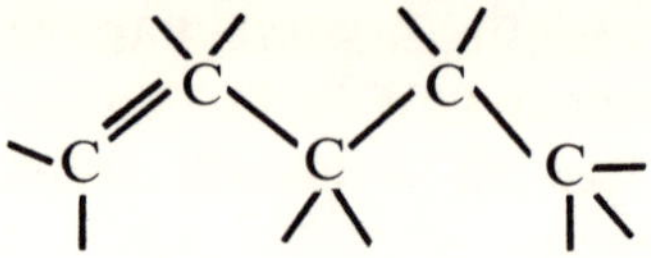

1-pentene

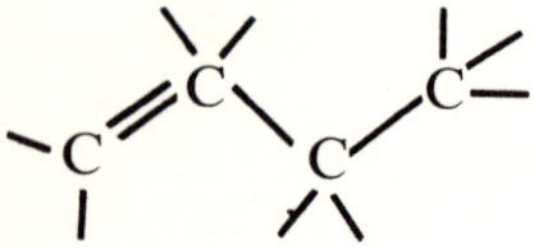

1-butene

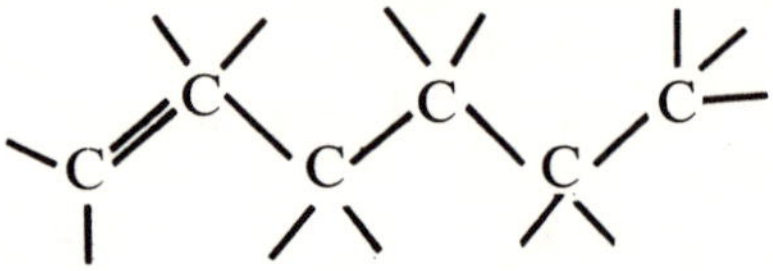

1-hexene

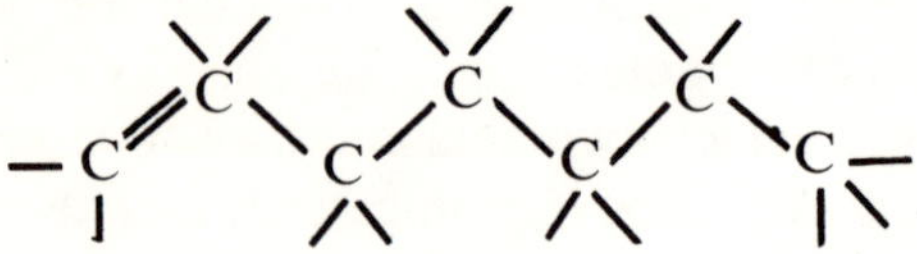

1-heptene

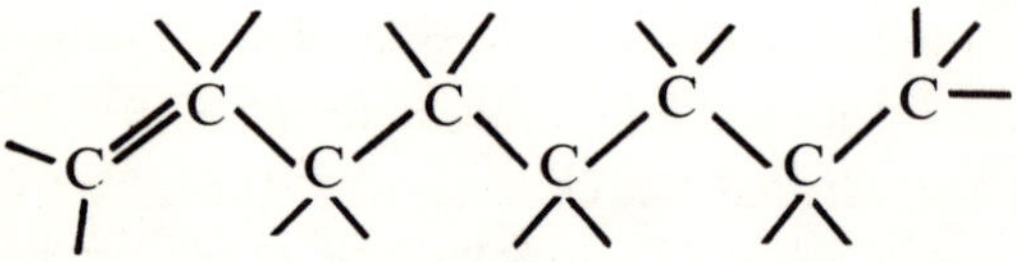

1-octene

FIG. 20  *Alkenes from $C_3$ to $C_8$. The number 1 in front of the name specifies that the double bond is between the first and second carbon atoms.*

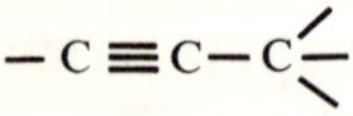

propyne 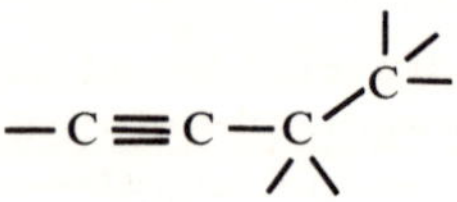      1-butyne 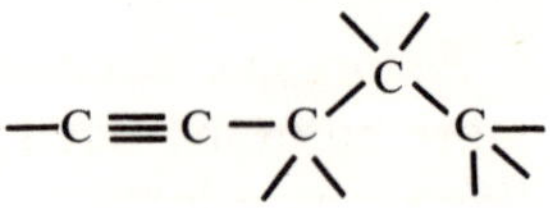      1-pentyne

FIG. 21  *Alkynes from $C_3$ to $C_5$.*

alkynes are important starting materials for a host of organic compounds. Ethyne, for example, goes into the manufacture of "the pill" — the oral contraceptive used by over 20,000,000 women.

Butadiene, $C_4H_6$, and isoprene (or 2-methyl-1,3-butadiene) are unsaturated hydrocarbons containing two double bonds. Molecules containing many double or triple bonds are said to be *polyunsaturated*. A molecule of polyisoprene (natural

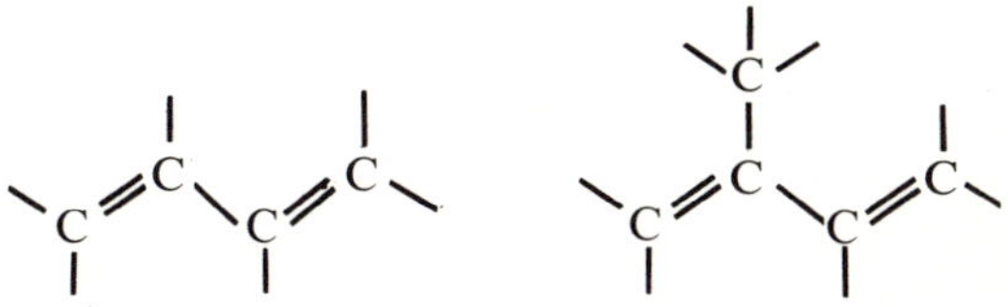

1,3-butadiene         isoprene (2-methyl-1,3-butadiene)

FIG. 22   *"Dienes"—hydrocarbons with two double bonds.*

rubber) is such a hydrocarbon, since it is composed of many isoprene molecules joined together. Polyisoprene is discussed further in Chapter VIII.

Like the alkanes, the unsaturated hydrocarbons may be either aliphatic (open-chain) or alicyclic (ring). The general formulas of the aliphatic alkenes and alkynes are $C_nH_{2n}$ and $C_nH_{2n-2}$, respectively. Obviously, the more multiple bonds there are between the carbon atoms, the fewer the number of hydrogen atoms in the molecule.

## Electrons on the Run: Aromatic Hydrocarbons

As organic chemists unraveled the structures of more and more naturally occurring molecules, it became apparent that there was a large group of compounds with fragrant aromas whose molecules possessed a unique structural feature. These compounds are the aromatic hydrocarbons and their unique structural feature is the flat, hexagonally shaped ring of six carbon atoms known as the "benzene ring." Benzene, $C_6H_6$, is the simplest aromatic hydrocarbon. For many years, chemists represented the structure of the benzene molecule as having alternating single and double bonds. Considerable experimental evidence has been compiled to show that this is not actually correct. If the benzene ring did have alternating single and double bonds, the ring would be lopsided and not

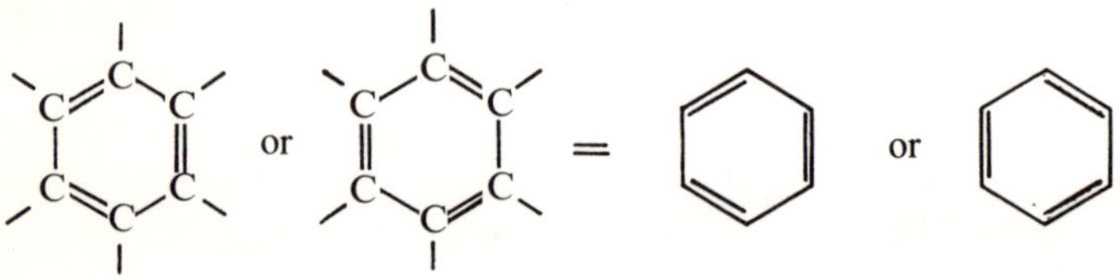

FIG. 23  *Classical representations of the benzene molecule.*

the regular hexagon which X-ray and electron-beam experiments reveal to be the case. The reason for the lopsidedness would be that the distance between the nuclei of adjacent carbon atoms varies according to the type of bond. Carbon atoms joined by a single bond are more distant than those joined by a double bond. The more bonds between two carbon atoms, the smaller the carbon-to-carbon distance. Comparing the bond lengths between the carbon atoms in ethane, ethene, and ethyne, we find ethane to have the largest and ethyne the smallest distance. In order to assume the shape of a regular hexagon, therefore, the six carbon atoms in the benzene ring must share equal numbers of electrons. Today, chemists still argue about exactly how this is accomplished, but they all agree that benzene does not have any double bonds. If it did, benzene would behave as an unsaturated hydrocarbon, which it does not. Ethene, for example, reacts rapidly with bromine at relatively low temperatures by a process called *addition*. The bromine atoms (there are two in each molecule) "add on" to the molecule of ethene to form an "addition compound" called 1,2-dibromoethane. The prefix "di-" refers to the two bromine atoms added to the ethene molecule, and the "1,2-" indicates that they are attached to different carbon atoms, numbered one and two. Benzene reacts with bromine in a far different manner. The reaction is comparatively slow and requires the input of more energy to get started. The process itself is called *substitution* because one of the two bromine atoms removes a hydrogen atom from the benzene ring to form hydrogen bromide while the other bromine atom takes the place of this hydrogen. The organic product of the reaction, monobromobenzene or just bromobenzene, for short, contains only one bromine atom per molecule. Alkanes, being saturated compounds, react by substitution, as does benzene.

$$C_6H_6 + Br_2 \rightarrow C_6H_5Br + HBr$$

What, then, might the structure of benzene be if it has no double bonds? If we assume that each of the six carbon atoms is bonded by a single bond to two other carbons and to one hydrogen each, then each carbon atom would be involved in the sharing of six electrons. This means that each carbon has utilized three of its four valence electrons in bond formation and still has one electron left over. We encountered a similar situation in the structure of graphite. As in the graphite crystal, the "extra" electrons in the benzene molecule are delocalized and move freely around the ring. Modern structural diagrams of the benzene molecule, therefore, show a circle within a hexagon instead of alternating single and double bonds to represent the electrons floating around the ring. Some chemists picture the benzene molecule as having three-electron bonds between adjacent pairs of carbons. Many chemists also view the benzene molecule from the orbital hybridization concept, in which the carbon atoms are

## 38 THE STUDENT CHEMIST EXPLORES ORGANIC COMPOUNDS

sp² hybridized. According to this concept, two of the three sp² orbitals overlap with the sp² orbitals of adjacent carbons while the third sp² orbital overlaps with the 1s orbital of the hydrogen atom, forming a total of three sigma bonds. The "extra" electron is in the unhybridized 2p orbital of carbon. The p orbitals, consisting of two lobes each, of adjacent

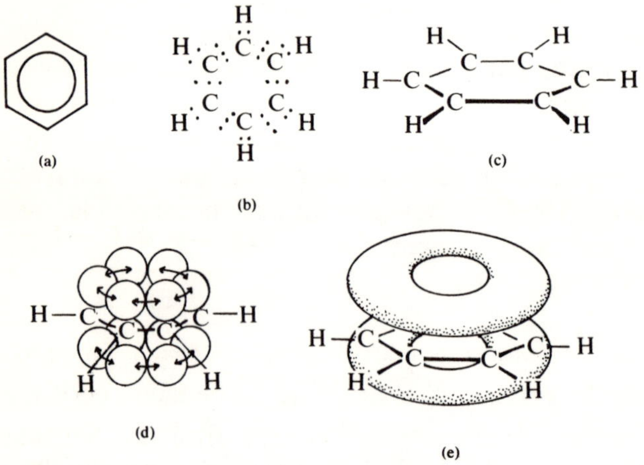

FIG. 24 *Modern representations of benzene.*
(a) *with circle designating delocalized electrons*
(b) *electron-dot formula showing three-electron bonds*
(c), (d), *and* (e) *orbital models:* (c) *shows orientation of carbon atoms with axes of their overlapping sp² orbitals (sigma bonds);* (d) *shows the overlapping p orbitals (pi bonds);* (e) *shows the doughnut-shaped orbital clouds above and below the ring resulting from the p-orbital overlap*

carbon atoms overlap side-to-side to form doughnut-shaped pi orbital clouds above and below the benzene ring. These clouds represent the probable location of the delocalized electrons.

There are several other molecules which fit the molecular formula $C_6H_6$, but these are chemically different from benzene. Compounds such as Dewar's benzene and prismane

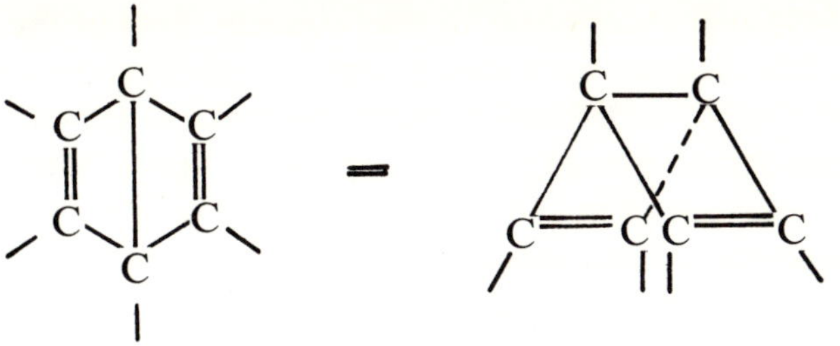

# Hydrocarbons: The Happy Union of Two Elements

**Dewar's benzene**

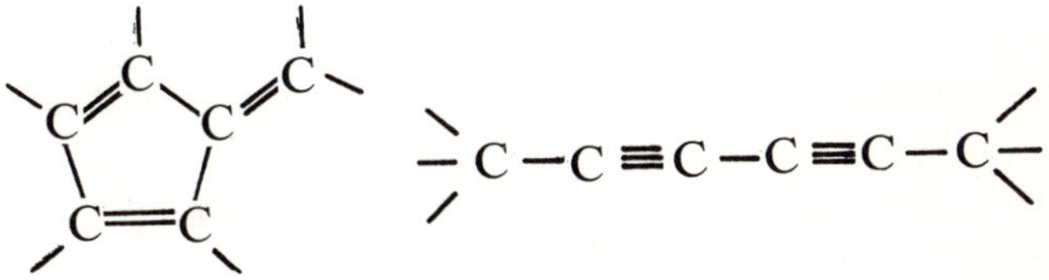

FIG. 25  *Non-aromatic molecules with a $C_6H_6$ formula. How many others can you draw?*

(see Chapter II) are not considered aromatic because they do not possess the special features of the benzene ring—namely, delocalized electrons or doughnut-shaped orbitals above and below the ring. There are, however, a great number of aromatic hydrocarbons. The name "aromatic" is really a misnomer since most aromatic compounds have, in fact, either unpleasant odors or no odor at all. Benzene itself, has a somewhat disagreeable aroma.

## Siamese Twins

Some aromatic hydrocarbons may have two or more "fused" rings, that is, rings with two carbon atoms in common. Napthalene, $C_{10}H_8$, once used as "moth balls," is the simplest such compound having two fused benzene rings. Those compounds with several fused rings are often carci-

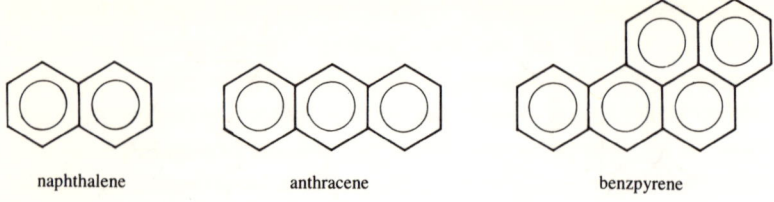

FIG. 26 *Polycyclic aromatic hydrocarbons: bicycles, tricycles, and polycycles.*

nogenic (cancer-producing). Benzpyrene is a carcinogenic agent found in cigarette smoke and in the air of every large city in the United States. It is formed during the burning of tobacco, coal, and such fuels as gasoline and diesel oil. It has been estimated that just by breathing the air in cities like New York or Detroit for one day a person inhales an amount of benzpyrene equal to that obtained by smoking 25 to 50 cigarettes.

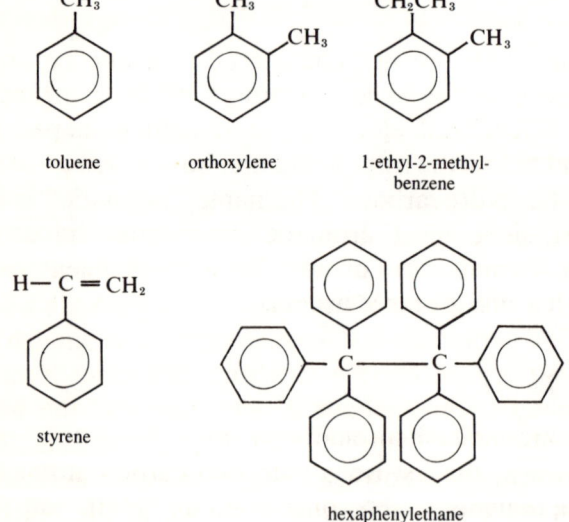

FIG. 27 *Examples of aromatic hydrocarbons with aliphatic and aromatic substituents.*

## Hydrocarbons: The Happy Union of Two Elements    41

Many other aromatic hydrocarbons consist of benzene rings with one or more attached aliphatic side groups. Toluene, $C_7H_8$, contains a benzene ring with an attached group containing one carbon and three hydrogen atoms. This "$CH_3$ group" is known as the *methyl* group because it is only one hydrogen removed from the alkane methane. Other *alkyl* groups are similarly obtained from corresponding members of the alkane series. The combination of the benzene ring with an assortment of alkyl groups gives rise to a remarkable torrent of aromatic hydrocarbons.

### With This Ring I Do Thee Wed

An unusual aromatic hydrocarbon synthesized recently is known as a catenane (from the Latin word *catena*, meaning "chain"). As the name implies, there are two independent rings in this compound which are interlocked as links in a chain. The molecular formula of one catenane is $C_{12}H_{12}$ (double $C_6H_6$). If you wonder how this strange structure got

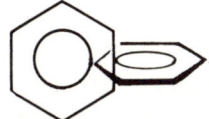

FIG. 28   *A catenane* — $C_{12}H_{22}$.

that way, here is the answer. During its formation, open chains of carbon atoms were made to bend in such a way that the terminal carbon atoms were brought close enough to form bonds, resulting in the ring structure. If two such molecules close around each other, or if one "threads" the hole of a ring-shaped molecule and then closes, catenanes are formed. Such a process is known as *ring closure* or *cyclization*. A simple example of cyclization is shown in the reaction between 1,3-dichloropropane and zinc. A zinc atom removes the two chlorine atoms, which are attached to the terminal

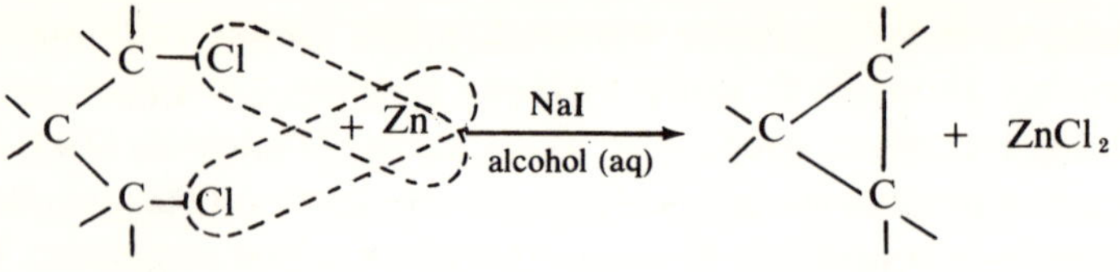

1,3-dichloropropane + Zn →(NaI, alcohol (aq)) cyclopropane + ZnCl$_2$

FIG. 29  *Cyclization: how to make a cycle from a chain.*

carbon atoms (numbered 1 and 3) in the dichloropropane molecule. As the chlorines are removed to form zinc chloride, the end carbon atoms bond together to form cyclopropane.

## What It Means To Be a Hydrocarbon

Whether aliphatic, alicyclic, or aromatic, hydrocarbons are virtually non-polar covalent compounds. The electrical forces by which the molecules attract one another are weak. These forces, called van der Waals' forces, depend upon the size and electron population of the molecule. The bigger hydrocarbon molecules, containing more electrons, are more easily distorted and "polarized." The more polar the molecules, the stronger the van der Waals' force. It is because the smaller hydrocarbons, such as ethane and ethene, are less easily polarized that their intermolecular forces of attraction are weakest, thereby enabling these compounds to exist as gases under ordinary conditions. Larger hydrocarbons, such as napthalene, have stronger van der Waals' forces and consequently are solids. Even the large hydrocarbons, however, will melt at much lower temperatures than most ionic solids of far lower molecular mass. The non-polar nature of hydrocarbons also accounts for their high solubility in non-polar solvents, such as carbon tetrachloride and benzene. They are usually scarcely soluble in water. The stronger forces of attraction between the polar water molecules make it very difficult for any hydrocarbon molecules to diffuse and intermingle with them. Since our bodies can utilize only soluble nutrients, those hydrocarbons that can power our cars, boats, and airplanes are useless to us as fuels. However, such important nutrients as edible fats and oils have molecules, huge segments of which are essentially hydrocarbon in nature. For example, more than three fourths of the mass of beef fat is due to its hydrocarbon portion. Therefore, every time we eat such fat-containing foods as butter, peanuts, milk, hamburger, and chocolate, our bodies must contend with a considerable amount of hydrocarbon-like molecules.

In this chapter we have examined some of the compounds that contain only two elements. We have seen that, although there are only two elements, the number of possible compounds is enormous. Furthermore, these compounds are of great importance to us. Now let us take the next logical step and broaden our exploration of the world of organic compounds. In Chapter V we will investigate those organic compounds that contain a third element in addition to carbon and hydrogen.

# V

# *Organic Relatives of $H_2O$, $H_2S$, and $NH_3$*

## *A Radical Substitution*

Binary compounds are compounds consisting of only two elements. Hydrocarbons are binary compounds. So are water ($H_2O$), hydrogen sulfide ($H_2S$), and ammonia ($NH_3$). Because they are inorganic, $H_2O$, $H_2S$, and $NH_3$ bear little resemblance to hydrocarbons other than that they also contain hydrogen. There are, however, a number of classes of organic compounds that are related to these three simple inorganic compounds. The organic "cousins" are derived by substituting one or more organic radicals (R), such as alkyl groups, for hydrogen atoms. Table 1 illustrates the resulting compounds. The R primes, R' and R", are used to signify that the radicals may or may not be identical. For example, R' may represent a methyl group, whereas R" may be a second methyl group, an ethyl group, or any other group.

## *One for the Road*

The organic counterparts of water are the alcohols. Alcohols contain one or more hydroxyl (OH) groups. The hydroxyl group is called a *functional group* because its presence in the molecule determines in effect how the molecule

Table 1
ORGANIC DERIVATIVES OF $H_2O$, $H_2S$, and $NH_3$

|  | H—O—H | H—S—H | H—N(H)—H |
|---|---|---|---|
| one hydrogen replaced | R—O—H alcohols | R—S—H mercaptans | R—N(H)—H primary amines[1] |
| two hydrogens replaced | R—O—R' ethers | R—S—R' sulfides | R—N(R')—H secondary amines |
| three hydrogens replaced | — | — | R—N(R')—R" tertiary amines |

The primary, secondary, and tertiary amines are collectively classified as amine.

behaves. It accounts for the similarity in behavior of different alcohols; it accounts for the differences among alcohols and other classes of compounds, including hydrocarbons. The hydroxyl group (or any functional group) acts as a single unit to specify a set of particular properties.

The two smallest alcohols are methanol or methyl alcohol, $CH_3OH$, and ethanol or ethyl alcohol, $C_2H_5OH$. The IUPAC[1] names for the alcohols are those ending in "ol." This system of nomenclature uses the parent alkane name and the suffix "ol" in place of "ane." These names, however, belie the close relationship between alcohols and water. Methanol and ethanol are very similar to water in many ways. Like water, these alcohols are polar, exhibit hydrogen bonding, can dissolve certain salts, and react with sodium metal to form a base and hydrogen gas. In fact, they are both liquids that are miscible[2] in water. However, as we consider larger

---

[1] The letters IUPAC stand for the International Union of Pure and Applied Chemistry.
[2] Miscible liquids are liquids that have unlimited solubility in one another.

# Organic Relatives of $H_2O$, $H_2S$, and $NH_3$

alcohol molecules, with bigger organic radicals, the resemblance to water correspondingly decreases. The organic radical portion assumes a more effective role in larger alcohols than in smaller ones. Much the same can be said for the organic counterparts of hydrogen sulfide and ammonia. In general, when organic radicals are substituted for hydrogen atoms in simple inorganic molecules, some original properties will be retained and some new properties will be introduced, how many of each depending on the size and complexity of the radical.

Many alcohols are *monohydric,* that is, they contain only one hydroxyl group. Dihydric alcohols and trihydric alcohols are also common. Ethanediol or ethylene glycol, a dihydric alcohol, is the chief ingredient of permanent antifreeze for automobiles. Propanetriol, commonly called glycerine or glycerol, is a syrupy, sweet-tasting liquid used in the production of skin lotions, candies, and, of course, nitroglycerine. Glycerine is a trihydric alcohol, its three hydroxyl groups explaining its marked affinity for water molecules.

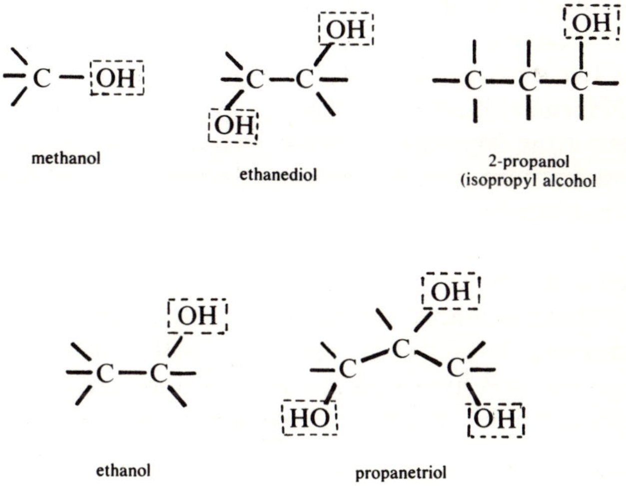

FIG. 30 *Some common alcohols.*

The best known alcohol is ethanol or ethyl alcohol. Man has long known that when certain fruit juices are allowed to stand for some time they ferment. This fermentation is a chemical change caused by the action of enzymes (catalysts) of yeast cells on sugars present in the juice. The sugars undergo a series of changes and are ultimately converted to ethanol and carbon dioxide. This process is the basis for the alcoholic beverage industry.

$$C_6H_{12}O_6 \xrightarrow{\text{yeast enzymes}} 2C_2H_5OH + 2CO_2$$

Although the drinking of fermented juices may produce some degree of euphoria (a "high"), any alcohol, even ethanol, should be properly regarded as a poison. Ethanol may be considered the most widely used drug in the United States—despite its toxicity. Concentrations of as little as from 0.4 to 0.6 percent of ethanol in the blood is enough to cause deep anesthesia or even death. One pint of pure ethanol would kill most people if ingested rapidly. The trick in drinking an alcoholic beverage (if one feels he must) without ill effect is to allow sufficient time for the body to oxidize the alcohol. Even so, excessive drinking of ethanolic beverages over a long period of time can cause liver damage, loss of memory, and addiction. Methanol, also known as wood alcohol, is an extremely potent poison. It is often added to ethanol to "denature" the latter and thus make it less fit for drinking.

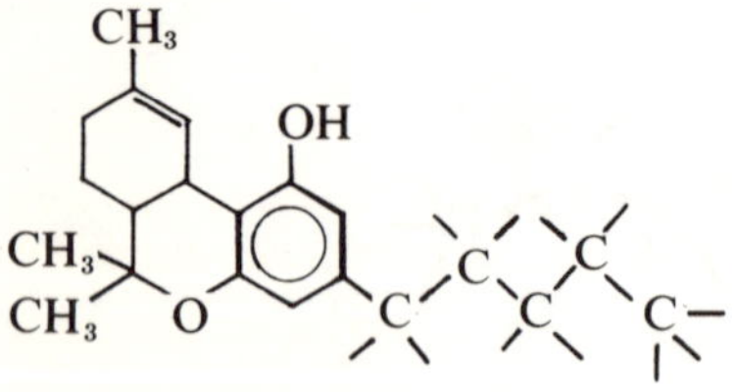

$\Delta^9$ tetrahydrocannabinol
(THE, "pot")

FIG. 31  $\Delta^9$ tetrahydrocannabinol (THC, "pot"). (The "$\Delta^9$" symbol indicates the position of the double bond.

## Organic Relatives of $H_2O$, $H_2S$, and $NH_3$ 49

Another well-known alcohol, tetrahydrocannabinol or THC, for short, is competing with ethanol for popularity among drugs. It, too, has been used by man for a long, long time. THC is the chemist's name for the active ingredient in marijuana. Research results have shown that THC has a more pronounced intoxicating effect than ethanol. Although not physically addictive, THC may develop in its users a psychological need for the drug. Many scientists agree that THC use is probably either less harmful or as harmful as smoking tobacco or drinking ethanolic liquors. In either case, not one of these activities has a very safe record.

Numerous biologically important compounds contain the hydroxyl group. These include sugars, sex hormones,

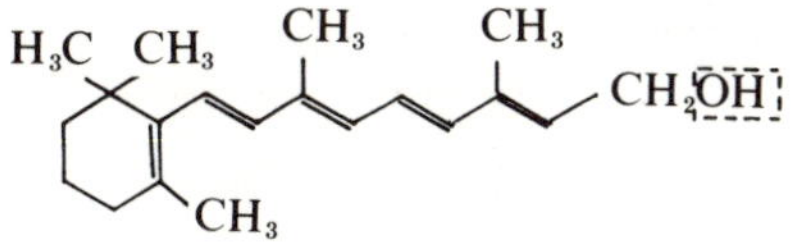

vitamin A, (retinol)

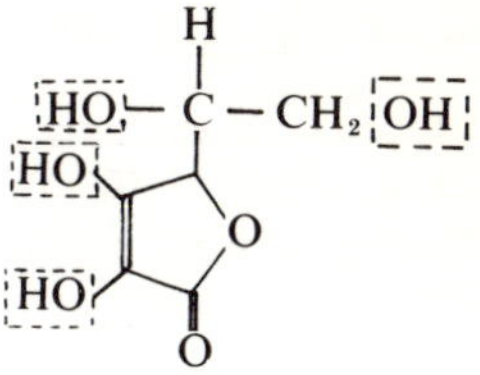

vitamin C (ascorbic acid)

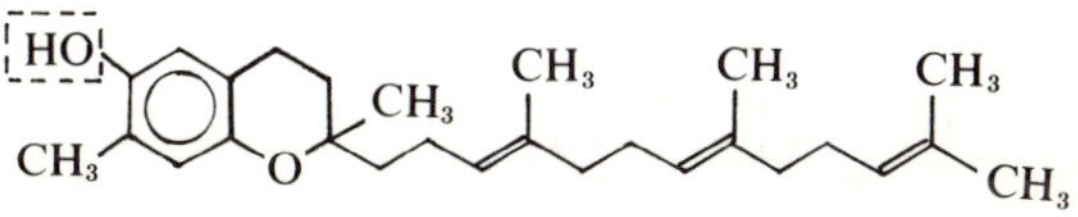

vitamin E (tocopherol)

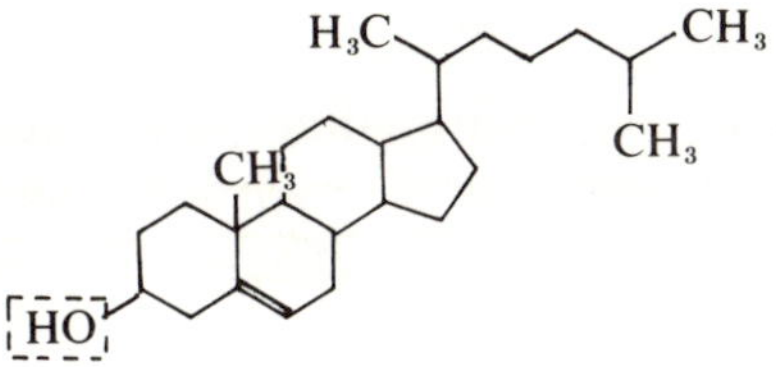

cholesterol

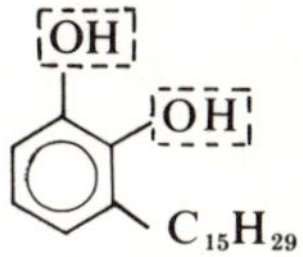

urushiol

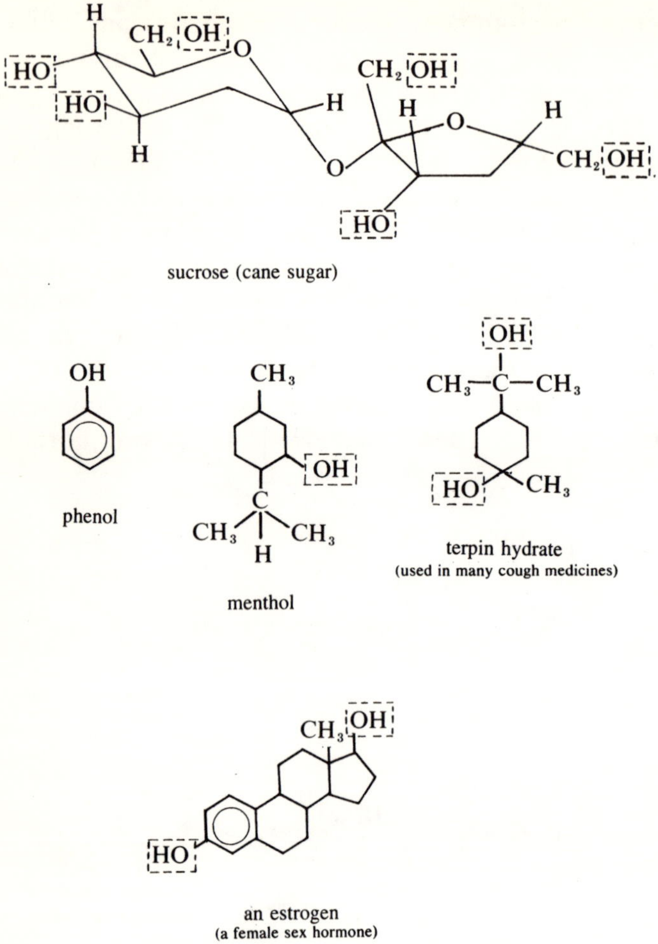

sucrose (cane sugar)

phenol

menthol

terpin hydrate
(used in many cough medicines)

an estrogen
(a female sex hormone)

FIG. 32 *Alcohols of biological interest.*

cholesterol, several vitamins, and adrenaline. Urushiol, a dihydric aromatic alcohol with a long hydrocarbon-like chain, is responsible for the unpleasant effects of poison ivy on our skin.

Fatty alcohols are usually monohydric alcohols corre-

## Organic Relatives of $H_2O$, $H_2S$, and $NH_3$

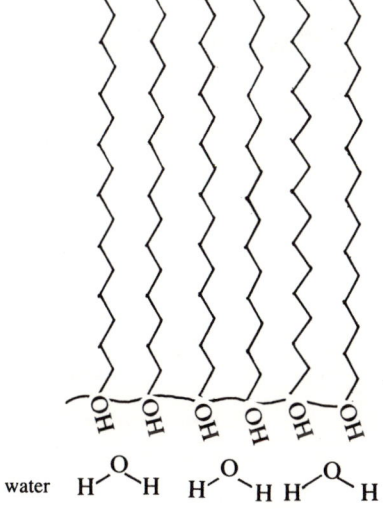

FIG. 33 *Fatty alcohols.*

sponding to the general formula R—OH, where R is a long, straight hydrocarbon-like group. Thus, $C_{18}H_{38}$, an alkane, becomes $C_{18}H_{37}OH$, stearyl alcohol, and $C_{18}H_{36}$, an alkene, becomes $C_{18}H_{35}OH$, oleyl alcohol. For some reason, fatty alcohols usually contain an even number of carbon atoms. Very small quantities of fatty alcohols with from sixteen to eighteen carbon atoms each are added to our lakes and reservoirs. Because these alcohols are so big, they are practically insoluble in water. Only the hydroxyl end is able to penetrate the surface of the water. Since their non-polar hydrocarbon segments stick straight up out of the water, these alcohol molecules in effect cover the surface of the lake or reservoir, reducing evaporation and conserving large quantities of water.

Cholesterol, the notorious substance that deposits on artery walls and contributes to heart attacks, is the kind of alcohol called a *sterol*. Sterols, in turn, belong to a larger group of compounds called steroids, which include the sex hormones, adreno-cortical hormones, and bile acids.

If both hydrogens in water are replaced by organic radicals, the compounds that result are called *ethers*. Ethers contain

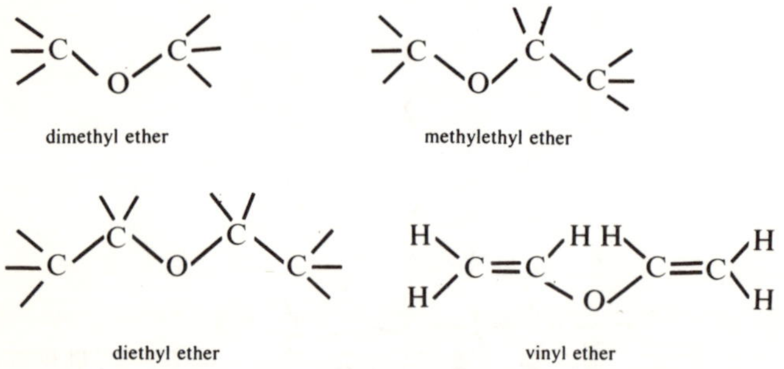

FIG. 34 *Some simple ethers.*

the characteristic "ether linkage," in which two carbon atoms are bonded to the same oxygen atom. Diethyl ether ($C_2H_5$—O—$C_2H_5$), is well known for its anesthetic properties. Another ether, vinyl ether ($C_2H_3$—O—$C_2H_3$), is a faster-acting and more potent anesthetic, but it is not as safe. Ethers have a dangerous combination of properties, inflammability and volatility, which necessitates the utmost caution in their handling and use.

## Organic Relatives of $H_2O$, $H_2S$, and $NH_3$

### King of the Smells

Mercaptans (or thiols or thio alcohols) are similar to alcohols in structure and composition, the difference being that mercaptans have a function sulfhydryl (SH) group instead of a hydroxyl group. The simple substitution of a sulfur atom for an oxygen atom produces an astonishing difference in aroma. Alcohols, for example, are usually

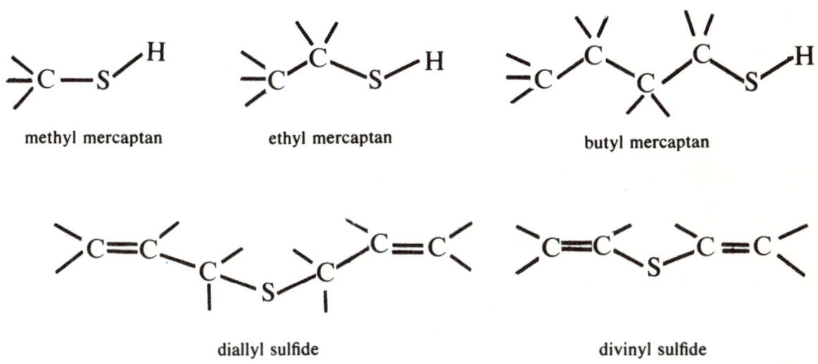

FIG. 35  *Simple sulfur compounds.*

odorless; mercaptans have possibly the most disagreeable of odors. In this respect, they are reminiscent of their inorganic counterpart, hydrogen sulfide, which has the odor of decaying eggs. Butyl mercaptan, $C_4H_9SH$, is chiefly responsible for the nauseating "eau-de-skunk." Larger mercaptans have stronger attractive forces, are less volatile, and therefore are less disagreeable in aroma. Many proteins, although they contain the SH group, are odorless thanks to their large size. It is difficult to think of a practical use of a mercaptan based on its noxious odor, but propyl mercaptan provides one example of such an application. Small quantities of this mercaptan are added to natural gas, which is itself odorless, to give aromatic warning of an open or leaky gas jet. Probably more than a few serious explosions have been prevented by using one of the best gas detectors known—the nose!

Diallyl sulfide and divinyl sulfide are thioethers, which resemble ethers but have a sulfur atom in place of the oxygen atom. The former sulfide is responsible for the taste and odor of garlic; the latter, for the taste and odor of onions. Thioethers may be thought of as organic sulfides, whereas ethers are organic oxides.

*Organic Bases*

What alcohols are to water and mercaptans are to hydrogen sulfide, amines are to ammonia. Replace one, two, or all three hydrogen atoms in ammonia and the result is an amine.

Amines are usually basic in nature (whereas alcohols and mercaptans are not) because of the pair of unbonded electrons on the nitrogen atom. In many of its reactions, the nitrogen atom in the amine accepts a proton ($H^+$) from water or an acid to form a positively charged ion analogous to the ammonium ion.

$NH_3 + H_2O \rightarrow NH_4^+ + OH^-$  $CH_3NH_2 + H_2O \rightarrow CH_3NH_3^+ + OH^-$
$NH_3 + HCl \rightarrow NH_4^+ + Cl^-$  $CH_3NH_2 + HCl \rightarrow CH_3NH_3^+ + Cl^-$

Because of their basic properties, amines are more soluble in acid solutions than in water. The more acidic the solution, the more soluble the amine. This simple fact has enormous health implications because proteins, the "building blocks of life," contain the functional amine group ($-NH_2$) and their solubility is largely determined by the pH (a measure of acidity) of the surrounding body fluid. Also, when the amine part of a protein adds on a proton, the shape of the protein molecule is altered. This is another important health consideration because proteins also comprise the body's enzymes and many of its hormones. Good health, therefore, requires that our body systems regulate ever so carefully the pH of our internal fluids.

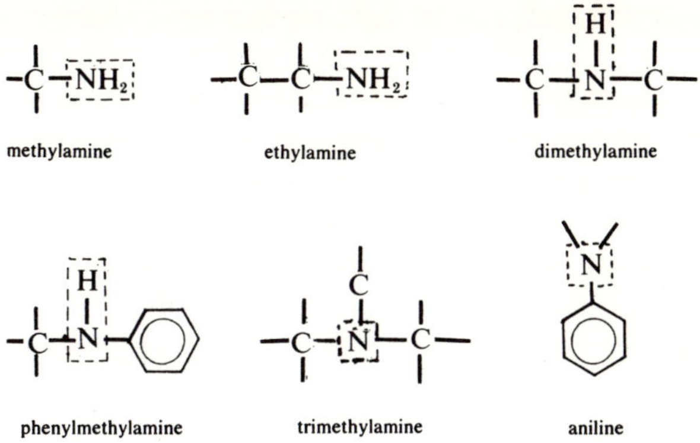

## Organic Relatives of $H_2O$, $H_2S$, and $NH_3$

FIG. 36 *Some typical amines:* (a), (b), *and* (f) *are primary amines;* (c) *and* (d) *are secondary amines;* (e) *is a tertiary amine.*

When meats or fish are left unrefrigerated for some length of time, the omnipresent bacteria secrete enzymes which catalyze the decomposition of proteins. The products of this fermentation are a number of simple amines with smelly odors and a more complex group of diamines (two $-NH_2$ groups) called *ptomaines*. You have undoubtedly heard of the dire consequences of ptomaine poisoning.[1] The names of the ptomaines, cadaverine and putrescine, could not be more appropriate.

There are numerous other amines capable of marked physiological effects, especially on the central nervous system. A group of synthetic amines, the *amphetamines*, are

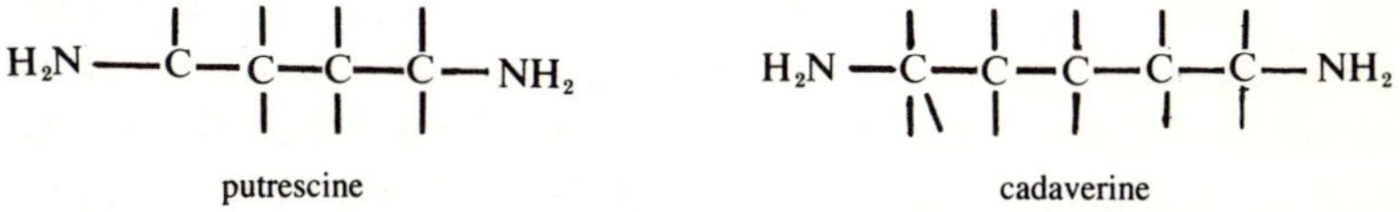

FIG. 37 *Two common ptomaines.*

---

[1] Actually, the ptomaines are not as poisonous as once believed. Other more complex chemicals produced by bacterial action have been found to be far more toxic. The term "ptomaine poisoning" has properly given way to the more accurate expression "food poisoning."

mind-altering drugs and are commonly called "speed," "uppers," and "pep pills." Amphetamines can decrease fatigue and elevate the user's mood by stimulating his central nervous system. In this sense, they mimic the action of the body's own psychic energizer, adrenaline (epinephrine), a structurally related amine. The amphetamines, under the trade names of Benzedrine, Dexedrine, and Methedrine, are easily manufactured and readily available. Some doctors,

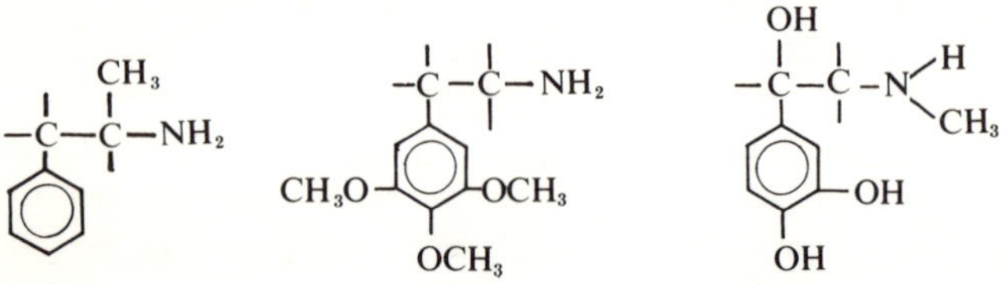


benzedrine mescaline adrenaline (epinephrine)

amphetamines

FIG. 38 *Amines that affect the central nervous system.*

unfortunately, prescribe these drugs all too freely, thus helping to make the use of amphetamines a major area of drug abuse.

In actuality, very few organic compounds lend themselves so neatly to an easy analogy with an inorganic relative. For simple, small organic compounds this is not too hard. Others, such as the heterocyclic amines, make analogies very difficult, if not impossible. The term "heterocyclic" implies a ring consisting of atoms of more than one element. In the case of a heterocyclic amine, this means that there are one or more nitrogen atoms incorporated in the ring along with the carbons! These compounds and others will be explored in the next chapter.

# VI

## *More Organic Compounds— For Life or Death*

### The Sour Ones

Why do vinegar, lemons, aspirin, rancid butter, and spoiled milk all have a sour taste? Probably because they all contain something in common which is responsible for the taste. As a student of chemistry, you recall that acids, such as hydrochloric and sulfuric acids, have sour tastes. This property is also typical of the organic acids present in each of the aforementioned substances. The type of organic acid encountered most frequently is the *carboxylic acid,* which contain a functional group called the carboxyl group, $-\overset{\overset{\displaystyle O}{\|}}{C}-OH$.

Just as alcohols may have more than one hydroxyl group in each molecule, there are monocarboxylic, dicarboxylic, and tricarboxylic acids containing one, two, or three carboxyl groups, respectively. The carboxylic acids, in many respects, resemble inorganic acids in their behavior, except that they are usually weaker. They form hydronium ions in water (usually in low concentration) and neutralize inorganic bases to form a salt and water.

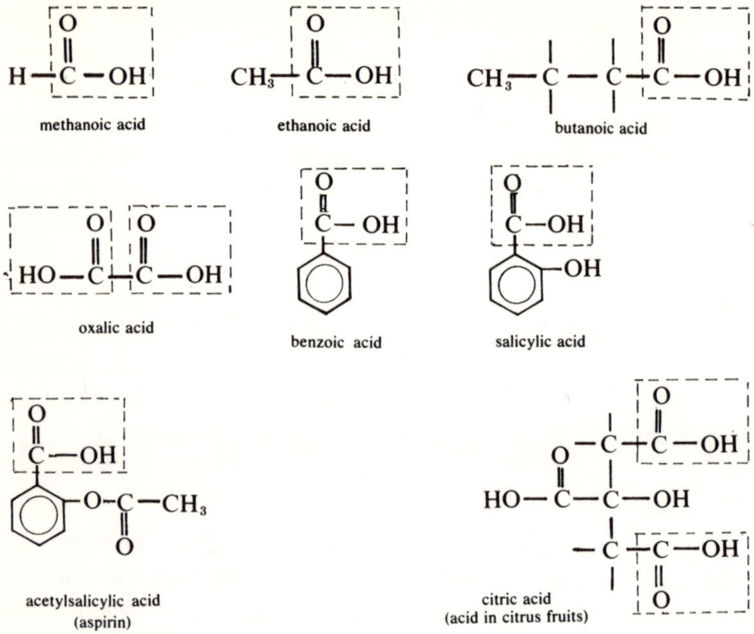

## 58 THE STUDENT CHEMIST EXPLORES ORGANIC COMPOUNDS

methanoic acid: H—C(=O)—OH

ethanoic acid: CH$_3$—C(=O)—OH

butanoic acid: CH$_3$—C—C—C(=O)—OH

oxalic acid: HO—C(=O)—C(=O)—OH

benzoic acid: C$_6$H$_5$—C(=O)—OH

salicylic acid: C$_6$H$_4$(OH)—C(=O)—OH

acetylsalicylic acid (aspirin)

citric acid (acid in citrus fruits)

FIG. 39  *Organic acids.*

$$CH_3-\overset{O}{\underset{\|}{C}}-OH + H_2O \rightarrow (CH_3-\overset{O}{\underset{\|}{C}}-O)^- + H_3O^+$$
acetic acid → acetate ion + hydronium ion

$$CH_3-\overset{O}{\underset{\|}{C}}-OH + NaOH \rightarrow CH_3-\overset{O}{\underset{\|}{C}}-O)^- Na^+ + H_2O$$
sodium acetate (an organic salt)

The simplest organic acid is formic acid. According to the IUPAC system of naming carboxylic acids, the name of the corresponding alkane with the same number of carbon atoms is used with the suffix "oic" replacing the "ane." Thus, formic acid, having one carbon (as does methane), becomes methanoic acid. The stings of many insects, including bees

## More Organic Compounds—For Life or Death

and wasps, hurt and cause swelling as a result of the injection of formic acid (along with other poisons) into our sensitive bodies. Vinegar is a dilute solution of acetic or ethanoic acid. This common acid is obtained by fermenting sugars and starches with bacterial enzymes. Louis Pasteur, the famous biochemist, explained a great mystery confronting French wine producers in the nineteenth century. The wine, it seemed, was developing an unpleasant sour taste. Pasteur showed that the ethanol in the wine, itself a product of fermentation by yeast enzymes, had continued to ferment in the presence of bacterial enzymes and atmospheric oxygen to form ethanoic acid, which caused the sour taste. Once the cause was known, the cure was relatively simple.

$$\text{glucose} \xrightarrow{\text{yeast enzymes}} \text{ethanol} \xrightarrow[\text{oxygen}]{\text{bacterial enzymes}} \text{ethanoic acid}$$

Similar reactions take place in the souring of milk and butter, resulting in the formation of lactic acid in milk and butyric or butanoic acid in butter. The similarity in the names "butter" and "butyric" is no coincidence. We have already seen examples of trivial, nonscientific names of compounds describing some aspect of its use, formation, structure, or properties. Stearic acid, for example, is obtained from the fat of a steer.

Aspirin, which is sold without prescription, is not now considered to be as harmless as once thought. The active component of aspirin, acetylsalicylic acid, is known to cause acid indigestion, internal bleeding, and, in some cases, a serious allergic reaction. So-called buffered aspirins have approximately the same degree of acidity as unbuffered ones —so why spend money needlessly on buffering ingredients? Aspirin, if needed, should be taken along with milk to help prevent irritation of cells lining the stomach. Oxalic acid, a dicarboxylic acid, has a high affinity for mercury. It combines with mercury to form insoluble mercury (II) oxalate.

Spinach and rhubarb are relatively high in oxalic acid content. If you are concerned about eating fish that might be contaminated with mercury, then do as Popeye did—eat a lot of spinach.

Fatty acids, such as stearic acid ($C_{17}H_{35}COOH$) and oleic acid ($C_{17}H_{33}COOH$), are carboxylic acids produced in the

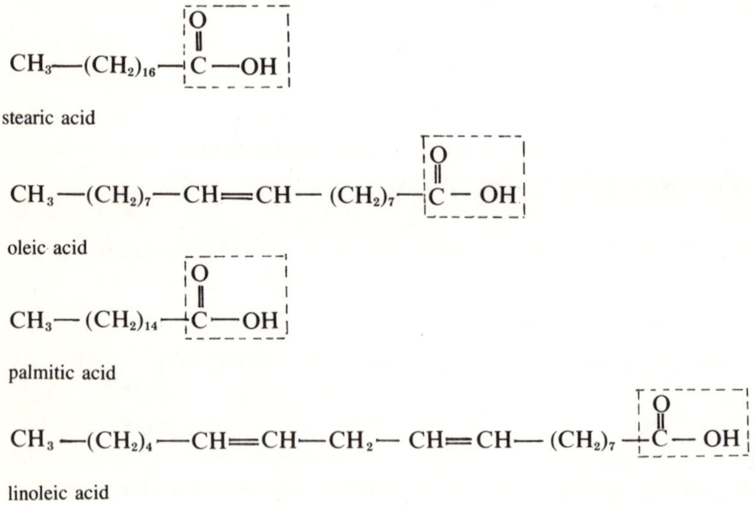

$CH_3-(CH_2)_{16}-\overset{\overset{O}{\|}}{C}-OH$

stearic acid

$CH_3-(CH_2)_7-CH=CH-(CH_2)_7-\overset{\overset{O}{\|}}{C}-OH$

oleic acid

$CH_3-(CH_2)_{14}-\overset{\overset{O}{\|}}{C}-OH$

palmitic acid

$CH_3-(CH_2)_4-CH=CH-CH_2-CH=CH-(CH_2)_7-\overset{\overset{O}{\|}}{C}-OH$

linoleic acid

FIG. 40 *Some common fatty acids.*

metabolism of fats. These acids have long, straight, hydrocarbon-like chains (as in the case of the fatty alcohols) with a single carboxyl group at one end. Because the major portion of the molecule resembles a hydrocarbon, fatty acids are not appreciably soluble in water. Our bodies, fortunately for us, have a mechanism for emulsifying fatty acids in the aqueous fluids (for example, blood). If this mechanism should falter for some reason, the resultant buildup of fatty acids in the body fluids could lead to clogged blood vessels and ultimately cause a "heart attack." The neutralization of a fatty acid with a strong base (such as sodium hydroxide) produces

## More Organic Compounds—For Life or Death

organic salts, which make excellent soaps. Sodium stearate is the chemical detergent used in bars of hand soap. Another organic salt, produced from glutamic acid (a dicarboxylic acid) and sodium hydroxide, is monosodium glutamate—well known for enhancing the flavor of meat:

$$\text{HO}-\underset{\text{glutamic acid}}{\overset{\overset{\displaystyle O}{\|}}{C}-CH_2-CH_2-\underset{\underset{\displaystyle NH_2}{|}}{C}-\overset{\overset{\displaystyle O}{\|}}{C}-OH} + \text{NaOH} \rightarrow$$

$$\text{HO}-\underset{\text{monosodium glutamate}}{\overset{\overset{\displaystyle O}{\|}}{C}-CH_2-CH_2-\underset{\underset{\displaystyle NH_2}{|}}{C}-\overset{\overset{\displaystyle O}{\|}}{C}-C-Na^+} + H_2O$$

$$\underset{\text{stearic acid}}{C_{17}H_{35}\overset{\overset{\displaystyle O}{\|}}{C}-OH} + \text{NaOH} \rightarrow \underset{\text{sodium stearate}}{C_{17}H_{35}\overset{\overset{\displaystyle O}{\|}}{C}-O^-\ Na^+} + H_2O$$

### From Carboxyl to Carbonyl

Closely related to the carboxylic acids, and to one another as well, are a number of classes of organic compounds known as *functional derivatives* of carboxylic acids. These derivatives are formed by replacing the —OH of $\overset{\overset{\displaystyle O}{\|}}{C}$—OH group with —Cl, —O—$\overset{\overset{\displaystyle O}{\|}}{C}$—R', —NH$_2$, or —OR', where the R' may be an aliphatic or aromatic radical. The resulting compounds are called acid chlorides, acid anhydrides, amides, and esters. The presence of the "C=O group," called the *carbonyl group*, in each of the derivatives makes them polar compounds and accounts for their chemical similarities. All but the amides have roughly the same boiling points, given molecules of comparable mass. The amides have relatively high boiling points because they are capable of forming strong

## Table 2
## ORGANIC COMPOUNDS CORRESPONDING TO THE FORMULA $R-\overset{\underset{\parallel}{O}}{C}-X$

| Functional group | Generic formula | Example | Common name (IUPAC name) |
|---|---|---|---|
| X = OH | $R-\overset{\underset{\parallel}{O}}{C}-OH$ | $CH_3-\overset{\underset{\parallel}{O}}{C}-OH$ | acetic acid (ethanoic acid) |
| X = Cl | $R-\overset{\underset{\parallel}{O}}{C}-Cl$ | $CH_3-\overset{\underset{\parallel}{O}}{C}-Cl$ | acetyl chloride (ethanoyl chloride) |
| $X = O-\overset{\underset{\parallel}{O}}{C}-R'$ | $R-\overset{\underset{\parallel}{O}}{C}-O-\overset{\underset{\parallel}{O}}{C}-R'$ | $CH_3-\overset{\underset{\parallel}{O}}{C}-O-\overset{\underset{\parallel}{O}}{C}-CH_3$ | acetic anhydride (ethanoic anhydride) |
| $X = NH_2$ | $R-\overset{\underset{\parallel}{O}}{C}-NH_2$ | $CH_3-\overset{\underset{\parallel}{O}}{C}-NH_2$ | acetamide (ethanamide) |
| $X = O-R'$ | $R-\overset{\underset{\parallel}{O}}{C}-O-R'$ | $CH_3-\overset{\underset{\parallel}{O}}{C}-O-CH_3$ | methyl acetate (methyl ethanoate) |

hydrogen bonds among their molecules. Acid chlorides and acid anhydrides react with water to form acids. These compounds usually have sharp, irritating odors. Volatile esters, on the other hand, often have pleasant, recognizable odors and are used in the preparation of many perfumes, artificial flavorings, and aromas. These esters are often the agents responsible for the fragrances of certain flowers and fruits. Isopentyl acetate, produced by the reaction between isopentyl alcohol and acetic acid, is actually "banana oil" and imparts to bananas a characteristic flavor and odor. Esters made from glycerine and fatty acids are fats, oils, and waxes.

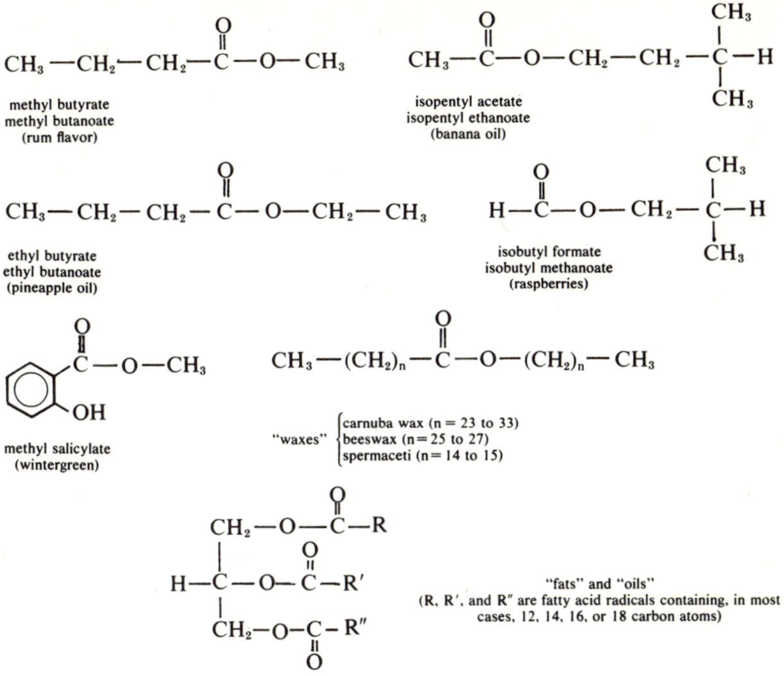

## More Organic Compounds – For Life or Death

$$CH_3-CH_2-CH_2-\overset{\overset{O}{\|}}{C}-O-CH_3$$
methyl butyrate
methyl butanoate
(rum flavor)

$$CH_3-\overset{\overset{O}{\|}}{C}-O-CH_2-CH_2-\overset{\overset{CH_3}{|}}{\underset{CH_3}{C}}-H$$
isopentyl acetate
isopentyl ethanoate
(banana oil)

$$CH_3-CH_2-CH_2-\overset{\overset{O}{\|}}{C}-O-CH_2-CH_3$$
ethyl butyrate
ethyl butanoate
(pineapple oil)

$$H-\overset{\overset{O}{\|}}{C}-O-CH_2-\overset{\overset{CH_3}{|}}{\underset{CH_3}{C}}-H$$
isobutyl formate
isobutyl methanoate
(raspberries)

methyl salicylate
(wintergreen)

$$CH_3-(CH_2)_n-\overset{\overset{O}{\|}}{C}-O-(CH_2)_n-CH_3$$
"waxes" { carnuba wax (n = 23 to 33)
beeswax (n = 25 to 27)
spermaceti (n = 14 to 15) }

$$CH_2-O-\overset{\overset{O}{\|}}{C}-R$$
$$H-\overset{|}{\underset{|}{C}}-O-\overset{\overset{O}{\|}}{C}-R'$$
$$CH_2-O-\underset{\underset{O}{\|}}{C}-R''$$

"fats" and "oils"
(R, R', and R" are fatty acid radicals containing, in most cases, 12, 14, 16, or 18 carbon atoms)

FIG. 41  *A few important esters.*

Waxes may also be produced from a fatty alcohol and a carboxylic acid. However it is produced, a wax has a long, hydrocarbon-like chain somewhere in its molecules.

Aldehydes and ketones also contain the carbonyl group, and their chemistry, too, is governed by the presence of this group. Aldehydes and ketones may be represented by the general formulas $R-\overset{\overset{O}{\|}}{C}-H$ and $R-\overset{\overset{O}{\|}}{C}-R'$, respectively. Formaldehyde is exceptional in that it contains two hydrogen atoms bonded to the carbonyl carbon atom. It has no radical. The IUPAC names for these carbonyl compounds are obtained by using the name of the corresponding alkane with the suffix changed to "-al" for aldehydes and "-one" for ketones. Formaldehyde, therefore, becomes methanal, and acetone, a ketone, becomes propanone. Acetaldehyde or ethanal is an intermediate substance formed during the metabolic conversion of ethanol to ethanoic (acetic) acid,

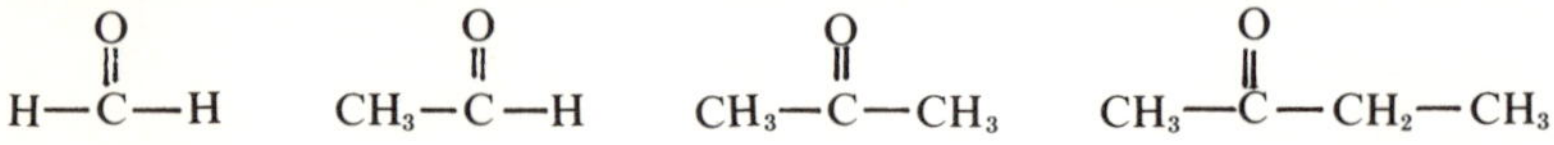

$$H-\underset{\underset{H}{\|}}{\overset{O}{\|}}-H \qquad CH_3-\underset{\|}{\overset{O}{\|}}-H \qquad CH_3-\underset{\|}{\overset{O}{\|}}-CH_3 \qquad CH_3-\underset{\|}{\overset{O}{\|}}-CH_2-CH_3$$

formaldehyde    acetaldehyde    acetone    methylethyl ketone
methanal    ethanal    propanone    butanone

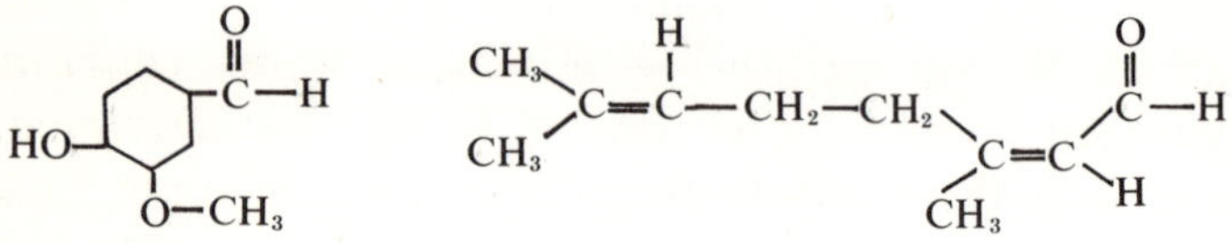

vanillin    citral (oil of lemon)

FIG. 42  *A few aldehydes and ketones.*

which takes place in the body of a person who drinks an alcoholic beverage. Although the complete mechanism is not known, it has been established that the ethanol changes first into ethanal by the action of one enzyme and that the ethanal is subsequently converted into ethanoic acid by the catalytic effect of a second enzyme. Antabuse is a drug used to discourage alcoholics from drinking by causing great discomfort if any ethanol is imbibed. What antabuse actually does is block the action of the second enzyme in the conversion of ethanol to ethanoic acid. In other words, it allows the first step in the conversion to proceed, but not the second. As a result, there is a buildup of ethanal, a highly toxic substance, which accounts for the unpleasant effect of antabuse in the presence of ethanol.

Aldehydes of low molecular mass are some of the chemical

## More Organic Compounds — For Life or Death

irritants in smog; they cause irritation of the eyes and mucous membranes even in light smog. These aldehydes are formed as a result of complex reactions between unburned hydrocarbons, primarily gasoline, in the atmosphere and various oxidizing agents present in smog, such as ozone and the nitrogen oxides. Propanone (acetone) and butanone (methyl ethyl ketone) are important industrial solvents. Butanone is commonly used in nail polish remover.

### Bring on the Halogens

The organic halides or organohalogens constitute another important class of organic compounds and may be considered derivatives of hydrocarbons with halogen atoms substituted for hydrogen atoms. Chloroform, $CHCl_3$, is a derivative of methane with three chlorine atoms in place of three hydrogens. It is, therefore, also named trichloromethane. By reacting methane with gaseous chlorine in varying proportions, it is possible to obtain the assorted chlorine derivatives of methane: $CH_3Cl$, $CH_2Cl_2$, $CHCl_3$, and $CCl_4$. Chloroform's use as an anesthetic has been curtailed because, like most chlorinated hydrocarbons, it is highly toxic to humans. Carbon tetrachloride or tetrachloromethane was actually used as a fire extinguisher and dry cleaning fluid for many years. It sometimes extinguished more than the fire.

Ethyl chloride (monochloroethane) is the local anesthetic administered as a spray to an injured athlete. Because of its high volatility, it evaporates rapidly, producing a freezing effect which minimizes swelling of the injured area. Vinyl chloride (monochloroethene) has been used as a propellant gas in aerosol spray cans. Sale of such aerosols for use in homes, hospitals, restaurants, and other enclosed areas was recently banned.

Another group of organohalogens are the freons, which contain both fluorine and chlorine. The freon designated as

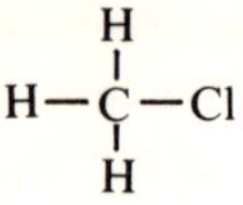

## 66 THE STUDENT CHEMIST EXPLORES ORGANIC COMPOUNDS

$$H-\underset{H}{\overset{H}{\underset{|}{C}}}-Cl \qquad H-\underset{Cl}{\overset{H}{\underset{|}{C}}}-Cl \qquad H-\underset{Cl}{\overset{Cl}{\underset{|}{C}}}-Cl$$

monochloro-
methane
methyl chloride

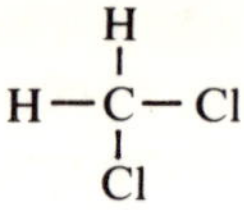

dichloro-
methane

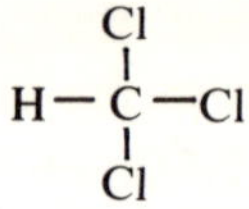

trichloro-
methane

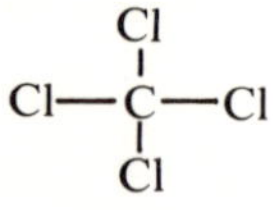

$$Cl-\underset{Cl}{\overset{Cl}{\underset{|}{C}}}-Cl \qquad H-\underset{\underset{H}{|}}{\overset{\overset{H}{|}}{C}}-\underset{\underset{H}{|}}{\overset{\overset{H}{|}}{C}}-Cl \qquad \underset{H}{\overset{H}{\diagdown}}C=C\underset{Cl}{\overset{H}{\diagup}}$$

tetrachloro-
methane

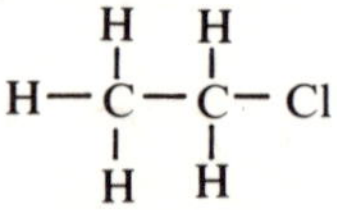

monochloro-
ethane
ethyl chloride

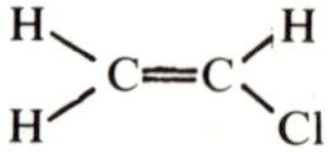

monochloroethene
vinyl chloride

FIG. 43 *Some simple organic chlorides.*

freon-12 is the refrigerant commonly used in air conditioners and refrigerators. It is also used to some extent as a propellant gas in aerosol spray cans. Supposedly it is nontoxic, but recently some research studies have raised some doubts about the safety of freon-12.

The idea of a man being completely immersed in a fluid for several hours or even days without suffering ill effects seems quite preposterous, doesn't it? Especially if the man has no special equipment to furnish oxygen to him. Yet, this may someday come to be. Mice have already been tested under such conditions and survived. The trick is that the oxygen needed for respiration was already dissolved in the fluid. The fluid must be special in that it must dissolve oxygen in fairly high concentrations. The fluid used was a fluorocarbon. Fluorocarbons are compounds containing carbon, fluorine, and, but not necessarily, hydrogen. The fluorocarbon, perfluorodecalin ($C_{10}F_{18}$), is made from the hydrocarbon decalin ($C_{10}H_{18}$) by substituting fluorine atoms for all the hydrogen atoms. Decalin consists of two six-membered carbon rings

## More Organic Compounds—For Life or Death

joined together along one edge. The rings are puckered to reduce strain in the molecule. Perfluordecalin has the ability to dissolve twice as much oxygen (and carbon dioxide) as whole blood. It is conceivable, therefore, that a fluorocarbon in conjunction with synthetic blood serum may someday be a substitute for whole blood. Scientists increasingly are becoming capable of altering not only our external environment but even the very fabric and substance of which our bodies are made.

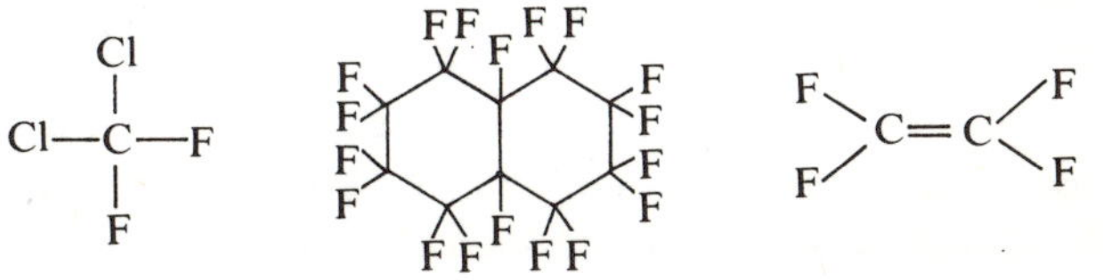

FIG. 44  *Organic fluorides of interest.*

freon-12    perfluorodecalin    tetrafluoroethene (used to synthesize teflon)

### The Longest War: Man vs. Insect

The longest war man has ever fought is the one waged against insects. Throughout history man and insect have competed for survival. The war is still a long way from being over and the end is not even in sight. Its short reproductive cycle and its ability to survive under conditions men could not tolerate make the insect a most formidable opponent, forcing man to use any kind of weapon he can devise. One of the first lines of attack (or defense) taken by man was the application of chemical insecticides. If man were to stop using insecticides today, probably 50 percent or more of his crops would soon be destroyed by insects. Such a large-scale destruction would cause havoc in a world already unable to feed all of its people. Thousands of people die of starvation each day even though insecticides are employed widely.

Among the first organic chemical insecticides tried were pyrethrin (from the pyrethrum flower) and nicotine (from tobacco). These were followed by the chlorinated hydrocarbons, the most famous being *di*chloro*di*phenyl*tri*-chloroethane[1] — or DDT, for short.[2]

Many stories about DDT's vital role in the war against insects have been told. In World War II, for instance, it was discovered that many German soldiers were sick with typhus as a result of being infested with lice. In fact, in previous wars more soldiers died from this cause than from actual battle wounds. The only reason the Allied soldiers did not suffer as much as the Germans did was because their bodies and clothing were treated with DDT. Many thought that DDT would rid mankind of all undesirable insect pests and the diseases they spread. It did, indeed, wipe out nearly whole populations of insect pests, such as flies and mosquitoes, in treated areas. However, this widespread use of DDT soon began to backfire and the insect menace became much more serious. Those insects that survived the first DDT onslaught soon reproduced and gave birth to new strains of insects that showed a greater resistance to DDT. These new strains required more massive doses of DDT for their demise. Again a few survived, which led to a still more resistant breed. This continued until strains of insects began to show no adverse affects at all from DDT. Moreover, the heavier doses of DDT also began to decimate some harmless species, not only of insects but of other animals for which it was not intended. DDT, which probably saved more lives than all the antibiotics combined, became a threat to the health of man as well and was considered an unacceptable risk. It was

---

[1] The "phenyl" group, $C_6H_5$, is the organic radical obtained by removing one hydrogen atom from the benzene molecule. It is, therefore, the simplest aromatic radical.

[2] If you think this name is long, you should see the chemical name of the protein called synthetane A with the formula $C_{1285}H_{250}N_{343}O_{375}S_8$. According to the *Guinness Book of World Records*, the chemical name for this compound contains 1,913 letters, making it the longest chemical name in existence.

## More Organic Compounds — For Life or Death

banned in the United States in 1972 and, soon after, the insect pests started to act up again. Some forest officials recently asked that the ban be lifted or else our forests would be destroyed. Many people would rather risk the consequences of insecticides than face diseases borne by insects and starvation as a result of ruined crops. What do you think?

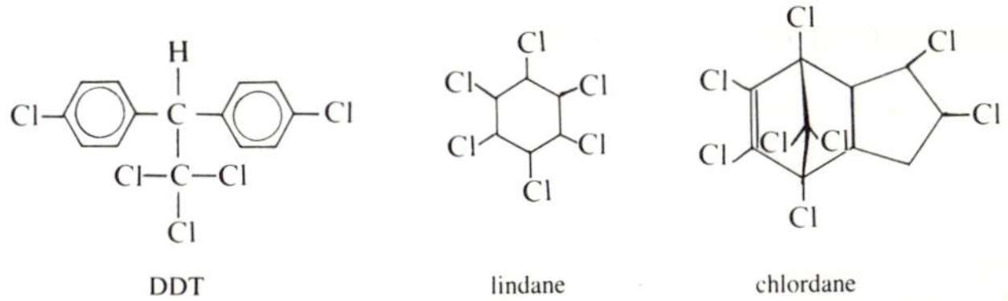

FIG. 45  *Organochlorine insecticides.*

How does DDT work? DDT is a nerve poison. Its molecules are non-polar and, therefore, tend to dissolve more readily in non-polar solvents, such as fats and oils. For this reason, the DDT that gets into the body of an animal concentrates in the fatty tissue and in the brain. The higher the organism (higher species in the food chain), the more fatty material it contains, and the greater the concentration of DDT. Scientists now are wondering if there are any other fat-soluble compounds like DDT that are increasing in concentration in living organisms, including you and me. If so, are any reaching levels that are considered dangerous to our well-being?

The fact that DDT molecules are fairly stable and fat-soluble, rather than water-soluble, presents another problem — they persist in the soil. Impervious to bacterial enzymes and immiscible with water, approximately half of the DDT sprayed can last in the soil for years. For this reason, the less-persistent organophosphates began to displace the chlorinated hydrocarbons as the chief chemical warfare agents.

The advantage of organophosphates, such as parathion and diazinon, is that they are *biodegradable* or *soft pesticides*. They do not remain long in the soil. Up to the present, insects have not shown any tendency toward developing a natural resistance to them. For example, 97 percent of malathion disappears in eight days. The organophosphates

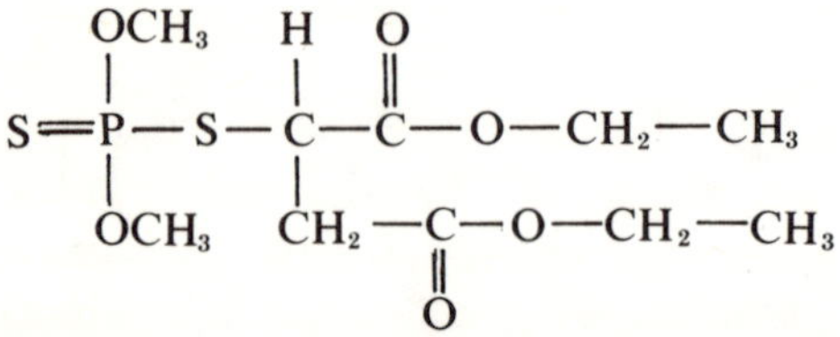

malathion

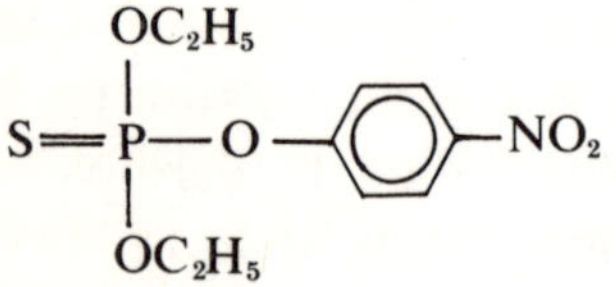

parathion

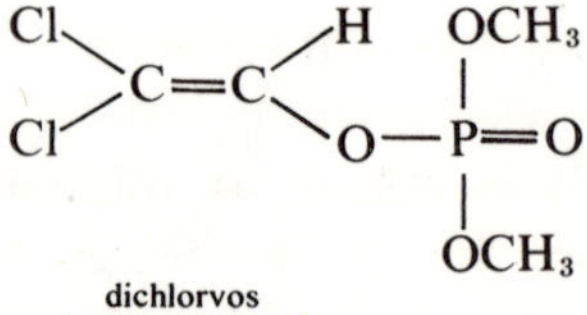

dichlorvos
(an ingredient of
Shell's "No Pest Strip")

FIG. 46 *Organophosphate insecticides.*

have a serious drawback, however, in that they are far more toxic to men and animals than DDT. Parathion[1] has caused more accidental human deaths than all other insecticides, including DDT, combined. DDT is relatively harmless to humans, as evidenced by the fact that some people (not I, for sure) have intentionally eaten large doses of DDT without any apparent ill effect. Do not try this experiment with DDT or any other insecticide! All the organophosphates have similar chemical structures, somewhat resembling those of

---

[1] The "thio" in the names parathion and malathion indicates the presence of sulfur atoms in the molecule.

# More Organic Compounds — For Life or Death

the "nerve gases" that have been developed for chemical warfare. They work on the same principle, but, unlike the nerve gases, the organophosphates are intended for insects. They act in such a way as to disrupt the chemical mechanism by which the body makes nerve conduction possible. Because of their toxicity, there have been attempts to ban some of the organophosphates.

## The Sexy Insecticides

A major disadvantage of chemical insecticides is the fact that they are somewhat indiscriminate. They may kill helpful insects, such as honeybees, which aid in pollination, as well as harmful ones. Although there are over 3,000,000 species of insects, only a few thousand are considered harmful. Why kill all insects if just a few are noxious pests? Clearly, another type of chemical control of insects is desirable. More and more attempts are being made in this direction. Chemosterilants, such as cytoxan and thio-TEPA, can cause sterility in the males of certain insect species. The sterile males are released after treatment with the chemosterilant, and, even though they can mate with females, no offspring will be produced since the eggs cannot be fertilized. The chemosterilants can, in this way, keep the species in check or, in some cases, lead to the eradication of the species entirely. Cytoxan is a heterocyclic compound, its unusual seven-membered ring containing atoms of oxygen, phosphoros, and nitrogen, along with carbon atoms. Atoms of

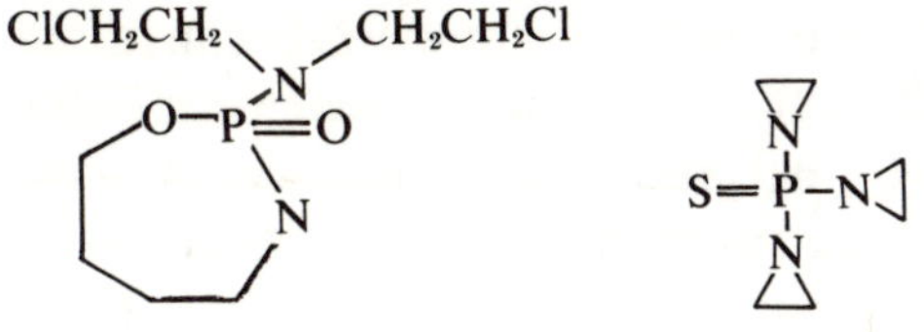

FIG. 47  Chemosterilants.

cytoxan

thio-TEPA

chlorine are found on the side chains, giving this molecule a total of six different elements.

Another method of controlling specific insect pests involves the use of compounds called *pheromones*. Pheromones are odorous chemicals which provide a means of communication among insects. They act as chemical messages from one insect to another regarding such matters as where to lay eggs, where to find food or a member of the opposite sex, removing dead bodies, and warning of intruders into a nest. The sexual pheromones, or sex attractants, are extremely powerful—just a few hundred molecules of a sex attractant emitted by a female may attract males of the species from as far away as two miles. Disparlure is the natural sex attractant of the female gypsy moth, a voracious insect that has caused great devastation of woodlands in the northeastern United States. If a minute quantity of disparlure (as little $10^{-12}$ gram) is placed into a trap, it will lure male moths from miles around. When a large number of males are lured near the trap, they may then be killed outright (by conventional insecticides) or made sterile (by chemosterilants). Another way disparlure can be used to control the gypsy moth is by simply spreading it around infested areas in order to confuse the stimulated males, who will not know which way to turn to find a female mate. It would be as though the male were surrounded by a bevy of gorgeous females who were totally invisible. What torture for the poor unsuspecting male! At any rate, no mating occurs, and the real females, if any were around, bear no offspring. The trouble with this particular method is that disparlure, although used in minute quantities, is hard to come by because the amount excreted by each female is also minute and the process of isolating the pheromone is laborious and expensive. A trouble with pheromones, in general, is the very specificity that we have also called an advantage. If there are 3,000 species of insect pests, at least as many pheromones

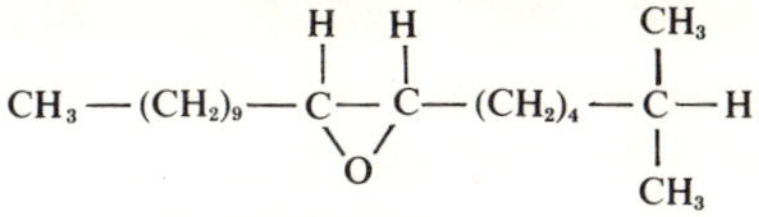

## More Organic Compounds — For Life or Death

$$CH_3-(CH_2)_9-\underset{\underset{O}{\diagdown\diagup}}{\overset{H}{\underset{|}{C}}}-\overset{H}{\underset{|}{C}}-(CH_2)_4-\overset{CH_3}{\underset{\underset{CH_3}{|}}{\overset{|}{C}}}-H$$

disparlure
(female gypsy moth sex attractant)

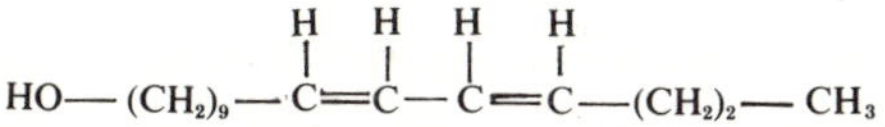

$$HO-(CH_2)_9-\overset{H}{\underset{|}{C}}=\overset{H}{\underset{|}{C}}-\overset{H}{\underset{|}{C}}=\overset{H}{\underset{|}{C}}-(CH_2)_2-CH_3$$

bombykol
(female silk moth sex attractant)

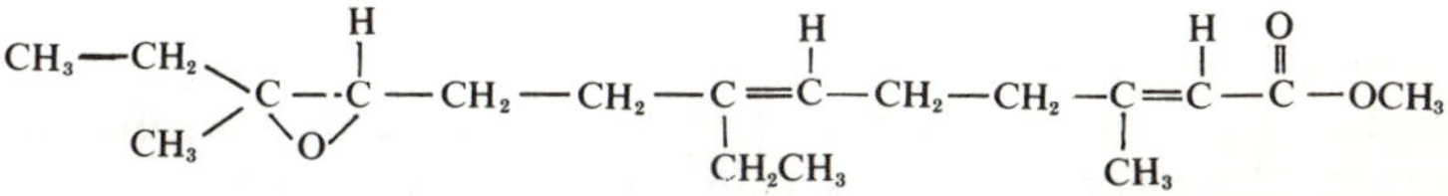

$$\underset{CH_3}{\overset{CH_3-CH_2}{\diagdown}}\underset{\underset{O}{\diagdown\diagup}}{\overset{H}{\underset{|}{C}}-\overset{}{\underset{|}{C}}}-CH_2-CH_2-\underset{\underset{CH_2CH_3}{|}}{\overset{H}{\overset{|}{C}}}=C-CH_2-CH_2-\underset{\underset{CH_3}{|}}{\overset{H}{\overset{|}{C}}}=\overset{O}{\overset{\|}{C}}-C-OCH_3$$

juvenile hormone
(cecropia moth)

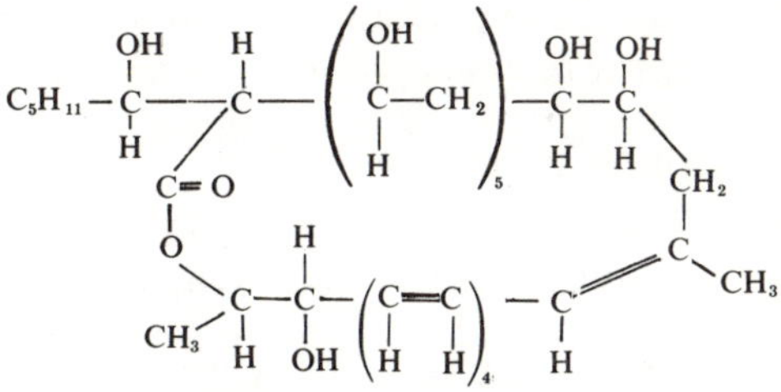

filipin

FIG. 48 *Various biochemical insect controls.*

will probably have to be identified and used, since what works on one species of insect may not have any effect on another. Even different species of ants, for example, do not all respond to the same pheromones.

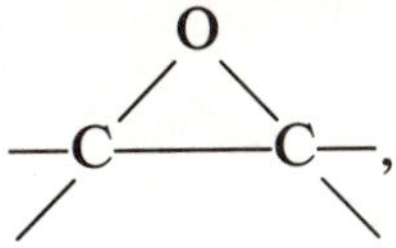

The molecule of disparlure contains a three-atom ring, one of which is an oxygen atom. Compounds having this particular ring, represented as $-C\overset{O}{\overset{\diagup\diagdown}{-\!\!-\!\!-}}C-$, are called epoxides or cyclic ethers. Unlike ordinary ethers, epoxides are extremely reactive due to the strain set up by the electron repulsion in the small epoxide ring.

Another epoxide called juvenile hormone represents still another line of attack that has many of the advantages and disadvantages of the pheromones. Juvenile hormone is one of the hormones secreted by an insect, a different one for each species, that regulates the growth and development of the young insect. If the insect is exposed to the juvenile hormone at the wrong time in its life cycle, then its development becomes abnormal. Exposed insects develop prematurely and die without ever being able to produce offspring. It is impossible for an insect to develop an immunity to these chemicals because they are a vital part of the insect's own normal development.

### *An Antibiotic That Kills Insects*

Cholesterol, which millions of people are convinced helps cause coronary heart disease, is a necessary component of the microscopic wall surrounding each cell in our bodies. Not only does it protect and insulate each cell in this capacity, but it also is needed to produce other chemicals vital to our existence. We do not need cholesterol in our diets, however, because our bodies can manufacture their own supply. Insects, too, require cholesterol, but unlike man cannot produce their own and so must have sufficient cholesterol in the diet.

It was recently discovered that some antibiotics, such as filipin, can prevent the absorption of cholesterol into the bloodstream, thus rendering the cholesterol useless. This property suggests how filipin could be used as an insecticide. Moth larvae that were fed with filipin were either killed or stunted. Similarly fed houseflies were either killed or made sterile. These hormones, like the pheromones, are extremely

## More Organic Compounds—For Life or Death

potent, one gram affecting up to a billion insects. The structure of filipin, as shown in Figure 48, reveals a 28-membered ring. Such large rings are not very common.

Chemists and other scientists have vowed to wage war against insect pests relentlessly until one or the other emerges victorious. Who knows—maybe we will win!

### Psychedelic Chemicals and Weed-Killers

Chemicals such as indole and skatole, which impart the characteristic odor of feces, certainly do not seem to be very psychedelic. However, the "indole ring" is recognized as a common component in the structure of a number of hallucinogenic or psychedelic drugs, including LSD, DMT, and serotonin. The indole ring is also a component of the structure of indoleacetic acid, a natural growth-regulating hormone in plants. Such hormones are used as herbicides

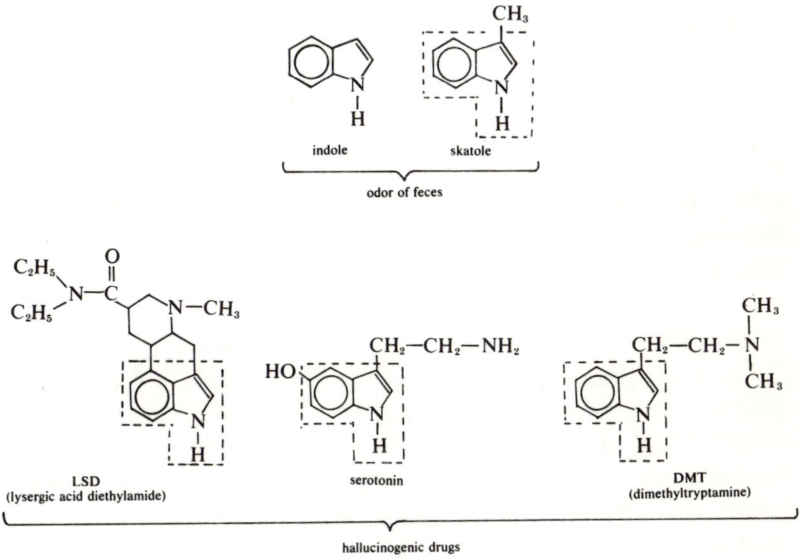

FIG. 49 *Compounds containing the indole ring structure.*

or weed-killers. They stimulate a plant to enlarge so rapidly that it literally grows itself to death. The growth hormone 2,4-D has been used for years as a weed-killer in lawns all over the country. Far more lethal is 2,4,5-T, which can kill practically any plant, small or large. It was used extensively in Vietnam to defoliate forests in which suspected enemy

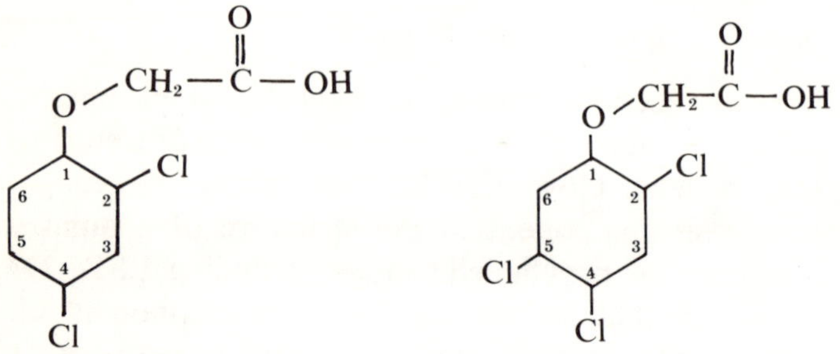

2,4-D
(2,4-dichlorophenoxyacetic acid)

2,4,5-T
(2,4,5-trichlorophenoxyacetic acid)

FIG. 50 *Organic herbicides. The numbers in the names designate the positions of the substituents on the ring.*

soldiers and their supplies and depots were hidden from aerial view. Recent studies showed that 2,4,5-T was probably responsible for the large percentage of birth defects in Vietnamese children born in the sprayed areas. Such chemicals are said to be *teratogenic,* in other words, "monster forming."

Another group of weed-killers destroys plants by blocking the normal photosynthetic process. The plants die because they are unable to produce glucose.

The alkaloids are heterocyclic compounds found in plants. They contain one or more nitrogen atoms. Many alkaloids, such as strychnine and nicotine, are very poisonous. Some, such as morphine, codeine, and heroin, are potent painkillers but are addictive. The alkaloid cocaine, also addictive, comes

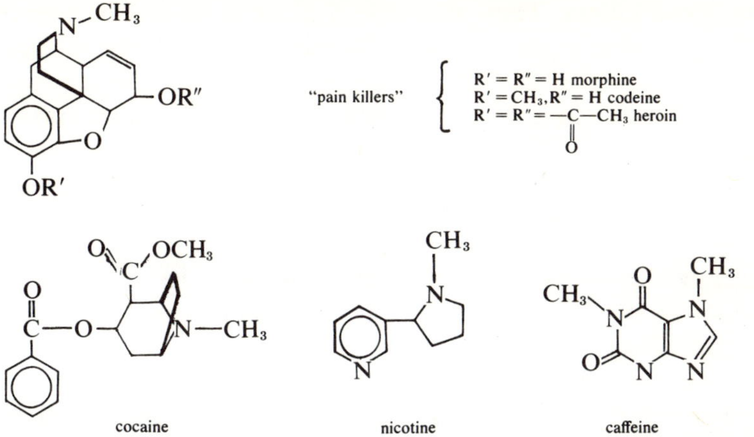

# More Organic Compounds—For Life or Death

FIG. 51 *Alkaloids.*

from the coca plant, from which cola drinks are made. Cola drinks, however, do not contain cocaine; but they do contain caffeine, another alkaloid. Caffeine is also found in tea, coffee, and cocoa. It acts as a stimulant of the spinal cord and brain and may be habit-forming. With this knowledge in mind, scientists are now considering the possibility that heavy coffee and tea drinkers, for example, might actually become addicted to these drinks—or, more specifically, to the caffeine in them. Results of investigations concerning this matter are not quite conclusive. You have probably heard the familiar coffee drinkers' lament that they feel sick and nervous until they get the next cup of coffee. Whether addictive or not, caffeine stimulates the secretion of gastric juice and may lead to instances of stomach distress.

Specific ring systems are also noted in a number of the alkaloids. These systems are shown in Figure 52.

Barbiturates, unlike alkaloids, are synthetic compounds. They are commonly prescribed by doctors as sedatives and tranquilizers—a very legitimate use if prescribed judiciously. However, too many people seek the "lazy" way to relax,

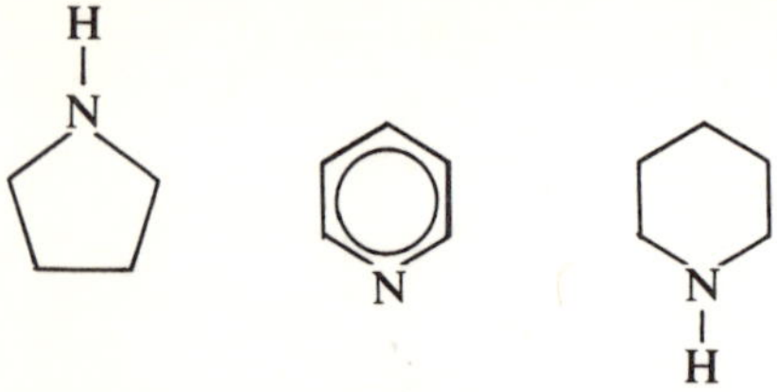

FIG. 52 Common ring structures containing a nitrogen atom.

pyrrolidine   pyridine   piperidine

relieve tensions, or get to sleep, and they have spawned a society of "pillpoppers." The barbiturates, which include such drugs as phenobarbital, chlorpromazine, and a host of "sleeping pills," produce the opposite effect of the amphetamines (Chapter V). If the amphetamines are the "uppers," the barbiturates are the "downers." Pity the poor soul who is addicted to both, depending on one to get to sleep at night and the other to arouse him in the morning. What a life!

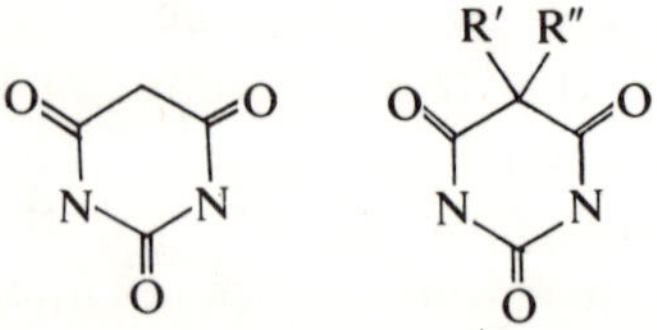

barbituric acid

generic formula for derivatives of barbituric acid

### THE TRANQUILIZERS

$R' = R'' = -CH_2CH_3$ barbital (Veronal$^R$)

$R' = -CH_2CH_3$, $R'' = -CH-(CH_2)_2-CH_3$ pentobarbital (Nembutal$^R$)
$\quad\quad\quad\quad\quad\quad\quad\quad\quad |$
$\quad\quad\quad\quad\quad\quad\quad\quad CH_3$

$R' = -CH_2CH_3$, $R'' = -\bigcirc$ phenobarbital (Luminal$^R$)

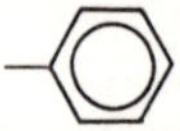

$R' = -CH_2CH=CH_2$, $R'' = -CH-(CH_2)_2CH_3$ secobarbital (Seconal$^R$)
$\quad\quad\quad\quad\quad\quad\quad\quad\quad\quad\quad |$
$\quad\quad\quad\quad\quad\quad\quad\quad\quad\quad CH_3$

FIG. 53 Barbiturates derived from barbituric acid.

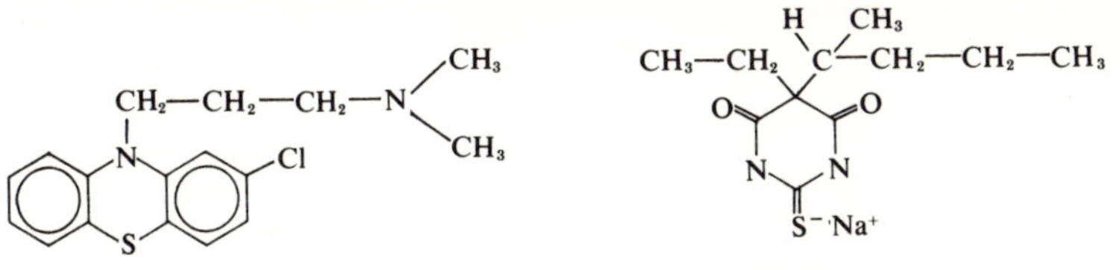

# More Organic Compounds — For Life or Death

**FIG. 54** *Other barbiturates. Notice that sodium pentothal is not far removed from the barbituric-acid derived barbiturates in structure.*

Just as the hallucinogenic drugs are derivatives of indole, the barbiturates are derivatives of barbituric acid.[1] Look at the formulas of these compounds again and note the similarities.

## They Are Not All Bad

In this chapter, we have explored only a small sampling of the ways in which a relatively few elements can be put together and give rise to a multitude of compounds which affect our lives in many significant ways. We have seen both the good and the bad.

Many other important compounds had to be omitted for lack of space. There are heterocyclic compounds, for example, that are essential to health. These include vitamin $B_1$, penicillin, hemoglobin, and the "base" components of nucleic acids — adenine, guanine, cytosine, and thymine. Organophosphates, too, are basic for life, as evidenced by the many phosphorus-containing enzymes, such as ATP, the phospholipids, vitamin B-12, and the nucleic acids.

---

[1] According to one story, barbituric acid is so named because it was allegedly discovered on St. Barbara's Day.

# VII

## Isomers: It's How You Put It All Together That Counts

*What It Means To Be an Isomer*

Compare the various properties of compounds A and B shown in the table below:

| Property | Boiling Point | Melting Point | Density | Solubility in Water | Reaction with Sodium |
|---|---|---|---|---|---|
| Compound A | −79°C | −117°C | 0.8 g/ml | complete | vigorous |
| Compound B | −24°C | −139°C | 2 g/ml | slight | none |

Apparently, compounds A and B are quite different. Yet, both compounds have the same molecular composition: two carbon atoms, six hydrogen atoms, and one oxygen atom. The formula $C_2H_6O$ could be used to represent either compound. This formula, however, does not in any way reveal the structure of the molecule nor does it help us ascertain which class of organic compound it represents. Given a set of models (ball and stick), you easily could determine that only two different structures are possible corresponding to $C_2H_6O$: one with the oxygen atom between a carbon and a hydrogen and the other with the oxygen atom between the two carbons. It is apparent that the chemical and physical

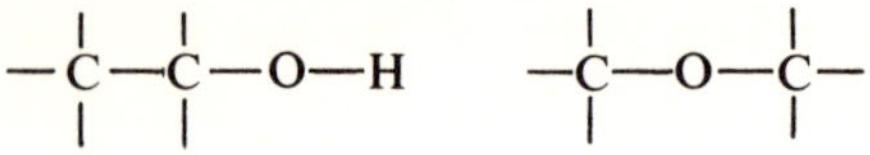

## 82 THE STUDENT CHEMIST EXPLORES ORGANIC COMPOUNDS

$$-\overset{|}{\underset{|}{C}}-\overset{|}{\underset{|}{C}}-O-H \qquad -\overset{|}{\underset{|}{C}}-O-\overset{|}{\underset{|}{C}}-$$

ethyl alcohol
ethanol
(compound A)

dimethyl ether
(compound B)

FIG. 55  *Two different molecules represented by* $C_2H_6O$.

behavior of a compound depends not only on its chemical composition but also on how its atoms are arranged within the molecule. The phenomenon by which two or more different compounds, with different sets of properties, can be represented by a single formula is called *isomerism*. Compounds which exhibit this phenomenon are called *isomers*. As we shall see in this chapter, there are many different kinds of isomers.

### Alkanes: One for All, All for One

The formula $C_8H_{18}$ may represent not just one compound but eighteen compounds, each identical in composition and structurally different from the others. Isomerism occurs among the alkanes having four or more carbon atoms per

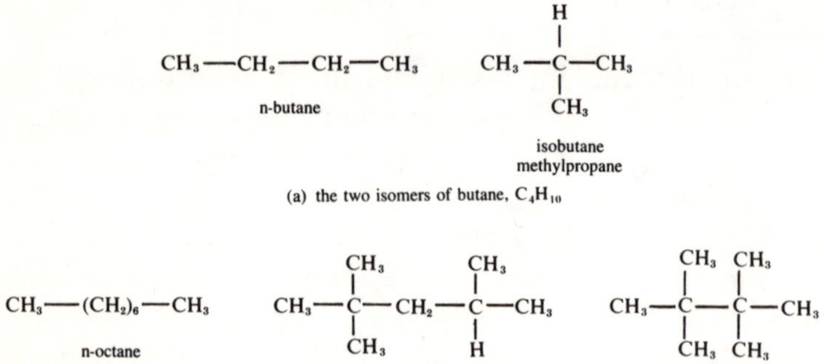

$$CH_3-CH_2-CH_2-CH_3 \qquad CH_3-\overset{H}{\underset{CH_3}{\overset{|}{C}}}-CH_3$$

n-butane

isobutane
methylpropane

(a) the two isomers of butane, $C_4H_{10}$

$$CH_3-(CH_2)_6-CH_3 \qquad CH_3-\overset{CH_3}{\underset{CH_3}{\overset{|}{C}}}-CH_2-\overset{CH_3}{\underset{H}{\overset{|}{C}}}-CH_3 \qquad CH_3-\overset{CH_3}{\underset{CH_3}{\overset{|}{C}}}-\overset{CH_3}{\underset{CH_3}{\overset{|}{C}}}-CH_3$$

n-octane

2,2,4-trimethylpentane
(isooctane)

tetramethylbutane

(b) three of the eighteen structural isomers of octane, $C_8H_{18}$; Can you draw the other fifteen structures?

FIG. 56  *Isomer structures.*

## Isomers: It's How You Put It All Together That Counts 83

molecule. Butane, $C_4H_{10}$, has two isomers. As the number of carbon atoms increases, the number of isomers increases rapidly: pentane ($C_5H_{12}$) = 3, hexane ($C_6H_{14}$) = 5, octane ($C_8H_{18}$) = 18, decane ($C_{10}H_{22}$) = 75, and eicosane ($C_{20}H_{42}$) = 366,319 isomers (not all have been made). An estimate based on a mathematical formula suggests that for the alkane formula $C_{40}H_{82}$ a total of 62,491,178,805,831 isomers can be fashioned.

### Alkenes: Double Your Pleasure with a Double Bond

The number of isomers of the alkane with four carbon atoms (butane) is only two; the number of isomers of the alkene with the same number of carbons (butene) is five. Butene must be capable, therefore, of forming types of isomers that butane cannot. The main reason for this has

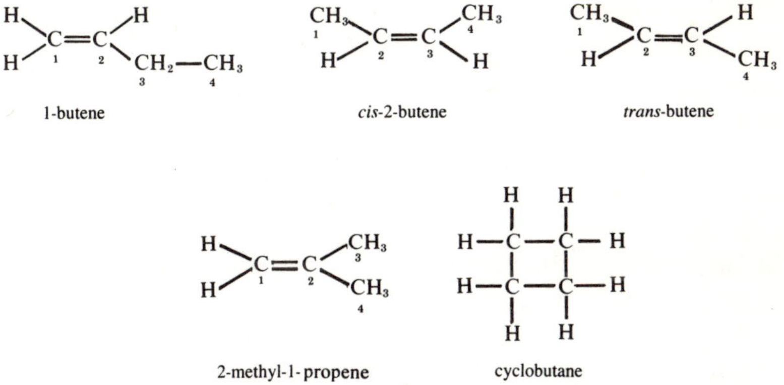

FIG. 57 *Isomers of* $C_4H_8$.

to do with the presence of the double bond in butene. The four carbon atoms in butene can be arranged in two different orientations, just as in butane, giving rise to two of the isomers. The other two isomers arise from the fact that the double bond can be situated in different places, that is,

84  THE STUDENT CHEMIST EXPLORES ORGANIC COMPOUNDS

between different pairs of carbon atoms. In 1-butene, the double bond is between the first and second carbon atoms (the carbons are numbered so as to give the first of the doubly bonded carbons the lowest number possible). In each 2-butene, the double bond is between the second and third carbon atoms. Careful inspection of the two 2-butene isomers reveals their fundamental difference. In each isomer, carbons numbered two and three have one methyl group apiece attached to it. In the isomer called cis-2-butene, the methyl groups are on the same side of the double bond; in the other, trans-2-butene, they are on opposite sides. If it were possible to rotate atoms freely about a double bond (as is possible with a single bond), there would be no "cis" and "trans" isomers, for one could be changed into the other by rotating the atoms around a doubly bonded carbon. All the butene isomers are similar in properties due to the lack

FIG. 58  *A chemical "no-no": rotation of atoms about a doubly bonded carbon is impossible.*

of any specific functional group. The double bond in this case may be considered as the functional group, since it determines the overall behavior of the compounds. The subtle variations in structure account for the slight, but observable differences in properties of the butenes. Cyclobutane, $C_4H_8$, is another isomer of butene (the fifth). It has most unusual properties because of its marked difference in structure. The pair of 2-butenes that differ only in their geometry or spatial arrangement of atoms, but not in overall structure or type of bond, are examples of *geometric* or *cis-trans isomerism*. Such isomers are only one type of a larger group of struc-

# Isomers: It's How You Put It All Together That Counts

turally similar isomers called *stereoisomers*. We will consider other types of stereoisomers later in this chapter. The isomers of butane are structurally unlike, each with a different assortment of bonds. They are called *structural isomers*.

## Cycloalkanes: The Nonconformers

The structures of the cycloalkanes containing four, five, and six carbon atoms are often represented as being two-dimensional, regular polygons (squares, pentagons, or hexagons). Benzene, you recall, has a flat, hexagonal ring

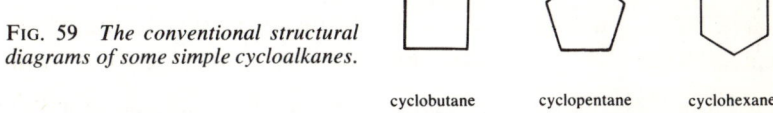

FIG. 59 *The conventional structural diagrams of some simple cycloalkanes.*

cyclobutane   cyclopentane   cyclohexane

structure. Such is not the case, however, in most cycloalkanes. In attempting to build a model of cyclobutane or cyclohexane with minimum ring strain, it is necessary to make the carbon-to-carbon bond angles as close to 109° as possible. Because the carbon atoms are all singly bonded, the atoms are $sp^3$ hybridized, and the geometry about each carbon should be that of a tetrahedron with angles of 109°. The molecules of cyclobutane and cyclohexane, therefore, have a bent, twisted, or puckered shape. The cyclopentane molecule is one of the few cycloalkanes with a nearly flat, molecular shape.

The molecules of cyclohexane exist as three different *conformations* or *conformational isomers*. Other orientations of the atoms would result in increased strain. The "chair" form is the most stable of the conformations, having the least amount of twisting and strain. The least stable is the "boat" conformation because the repulsion between the hydrogen atoms at the "flagpole" positions leads to a higher energy

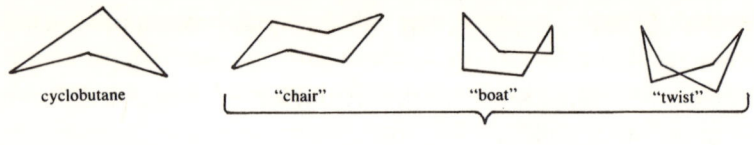

conformations of cyclohexane

FIG. 60

state. The names of the conformations correspond to the descriptions of the items one might construct with matchsticks.

The models of the chair conformation shown in Figure 61a show a number of dark lines and a number of light lines representing carbon-hydrogen bonds. The dark lines are said to be *axial* bonds, that is, bonds which extend vertically above and below the ring. The light lines (assuming the ring to be horizontal for convenience) are *equatorial* bonds which extend outward horizontally. The hydrogen atoms adjoined

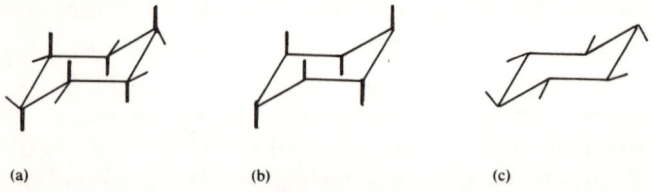

FIG. 61  *Axial and equatorial bonds in cyclohexane.*
(a) *shows all bonds in the chair conformer of cyclohexane*
(b) *shows only the axial bonds, and* (c) *shows only the equatorial bonds*

by these bonds are called axial and equatorial hydrogen atoms. That cycloalkanes, such as cyclohexane, have two different types of hydrogen atoms in the chair form suggests that isomers of these cycloalkanes can result by substituting other atoms or groups for these hydrogen atoms. If we substitute a methyl group for only one of the hydrogen atoms on a given carbon, the compound methylcyclohexane is

Isomers: It's How You Put It All Together That Counts  87

formed. There is only one methylcyclohexane, however, because of the ability of a chair form to convert to an oppositely facing chair form by flipping one end up and the other down, as shown in Figure 62.

Let us assume that before the conversion the methyl group was in an equatorial position. After the conversion, it has

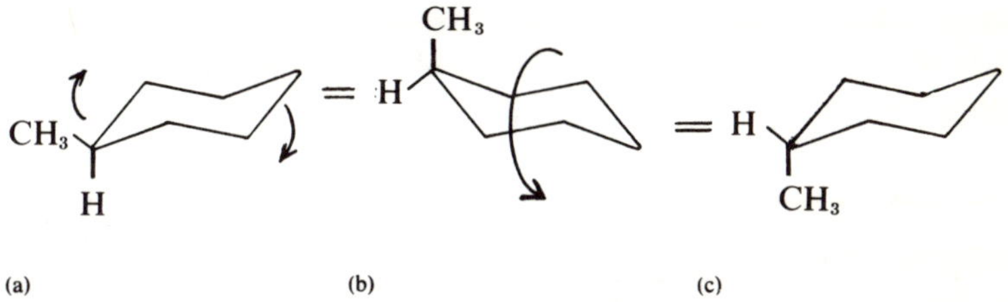

FIG. 62  Ring inversion in cyclohexane. In (a), the methyl group is in an equatorial position. Ring inversion occurs when the end of the chair conformer with the methyl group is flipped up and the opposite end is flipped down, producing the inverted chair form shown in (b). In (b), the methyl group is now in an axial position. By turning the inverted chair form (b) upside down, the carbon atoms in the ring are in a comparable orientation as those in (a). The only difference between (c) and (a) is the axial-equatorial nature of the methyl group. Actually, all equatorial positions became axial and vice versa. This inversion may occur a million times each second.

switched into what is now an axial position. If the converted chair form is turned upside down so that it faces the same direction as the original form, both forms will be identical except for the axial-equatorial interchange. In other words, the positions of the axial and equatorial atoms bonded to a given carbon atom are easily interchanged. Now, when more than one substituent is added, especially if the substituents are large enough, the ring may become so bulky that interconversion is made impossible. In this event it is possible to have isomers, depending on whether the attached groups are in equatorial or axial positions. For example, the two dichlorocyclohexanes shown in Figure 63 exist by virtue of the fact that one has both chlorines in the equatorial position

## 88 THE STUDENT CHEMIST EXPLORES ORGANIC COMPOUNDS

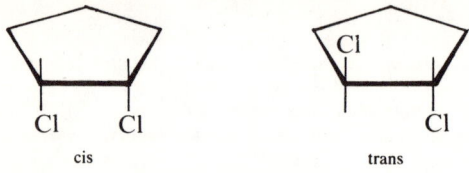

FIG. 63 *Isomers of 1,2-dichlorocyclohexane.*

and the other has one chlorine in an equatorial position and the other in an axial position.

The ring in cyclopentane is only slightly puckered, the bond angles being close to 108°. Assuming the molecule, therefore, to be nearly flat or planar, we find that it is possible to have cis-trans isomerism here as we did in the alkenes. In the

FIG. 64 *Isomers of 1,2-dichlorocyclopentane.*

cis-form of dichlorocyclopentane, there are two chlorine atoms on the same side of the ring or plane, whereas in trans-dichlorocyclopentane the two chlorines are on opposite sides. How many isomers of dichlorocyclopentane can you draw?

### Isomers, Rings, and Mothballs

Many aromatic compounds exhibit isomerism. For example, there are three possible dichlorobenzenes, which are made by substituting two chlorine atoms on the benzene ring. One system of naming these isomers uses the numbering scheme in which a particular carbon is assigned the number one and the other carbons are numbered consecutively in a clockwise or counterclockwise direction around the ring. The carbon atom which in the end result gives the lowest

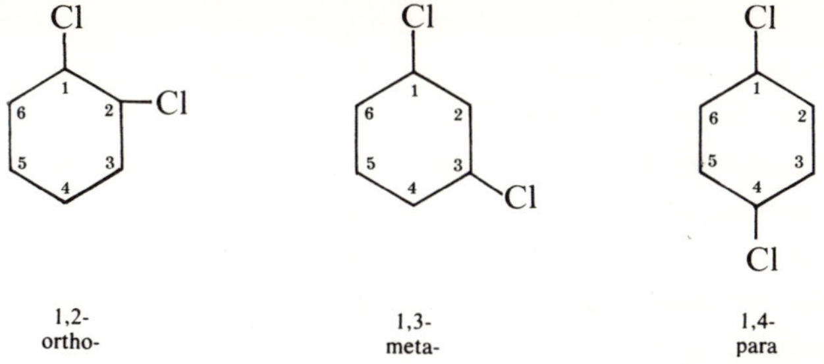

## Isomers: It's How You Put It All Together That Counts

[Structures: 1,2- ortho-; 1,3- meta-; 1,4- para-]

FIG. 65  *Isomers of dichlorobenzene.*

number on all the carbons is the one selected as number one. Another system uses the prefixes "ortho-" (adjacent carbons), "meta-" (one carbon between), and "para-" (opposite carbons) instead of numbers to indicate the carbons containing attached groups.

The fact that there are only three isomers of dichlorobenzene is further proof that benzene does not contain double bonds in the ring (see Chapter IV), for if it did there would be two isomers of 1,2-dichlorobenzene, as shown in Figure 66: one in which both chlorine atoms are attached to carbons,

FIG. 66  *Which is the real 1,2-dichlorobenzene?*

forming a double bond between them, and another when the carbons form a single bond between them. The fact is that there is only one structure for 1,2-dichlorobenzene.

### Looking-Glass Isomers

Are your two hands identical? Each one has the same number and kinds of fingers, but try to put your left hand in a right-hand glove and it does not fit. Another way to tell

90  THE STUDENT CHEMIST EXPLORES ORGANIC COMPOUNDS

that your two hands are not identical is that you cannot superimpose one hand over the other, palms facing the same way, and have your thumbs pointing in the same direction. To put it simply, your hands are not superimposable and the explanation for this simple fact of nature is that your two hands are mirror images of each other.

Can molecules have mirror images, too? How would mirror image molecules differ with respect to their chemical and physical properties? The answer to the first question depends on whether or not the molecule has an *asymmetric* carbon atom. An asymmetric carbon atom is one that has four different atoms or groups attached to it. Consider the central carbon in the molecule of the amino acid alanine, which in Figure 67 is drawn as a circle.

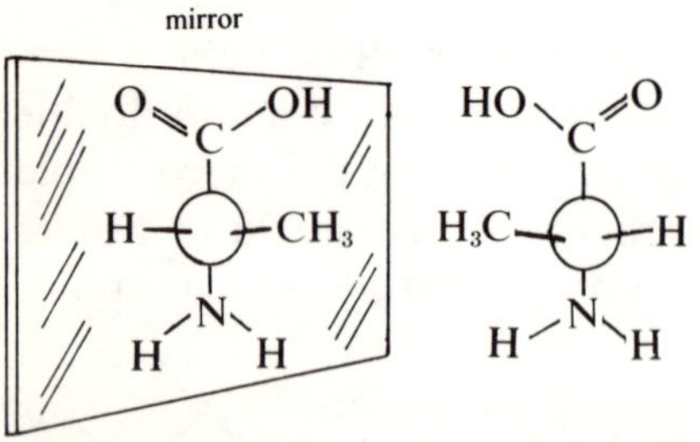

FIG. 67  *Mirror image of alanine,*

$$H_2N-\overset{\overset{\displaystyle CH_3}{|}}{C}H-COOH -the\ same\ or\ different?$$

There are, as you can see, four different chemical species attached to this carbon: one hydrogen atom, one methyl group (—$CH_3$), one amino group (—$NH_2$), and one carboxyl group (—$\overset{\overset{\displaystyle O}{\|}}{C}$—OH). The central carbon is, therefore, asymmetric. Now let us apply the superimposing test. Figure 67 will help us visualize the three-dimensional nature of the alanine molecule. Lines that extend into the circle (central atom) represent bonds directed out of the page toward you; lines that end at the circumference of the circle represent bonds directed away from you into the page. Of course, if

## Isomers: It's How You Put It All Together That Counts 91

you have a set of models handy you can construct these molecules very easily. If not, try turning the molecules in the page around mentally and see if they can be superimposed. Keep in mind the directions of the bonds—to be superimposed properly the atoms and their bonds must be identically aligned. You will find that these structures are different and cannot be superimposed. They are, as the figure shows, mirror images or, as they are often called, *enantiomers* or *enantiomeric forms*. Now try doing the same experiment with the structures of glycine, another amino acid, as shown in Figure 68. Are the structures superimposable? I think

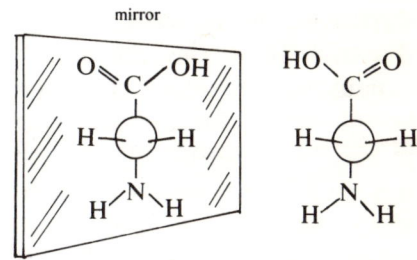

FIG. 68 *Mirror image of glycine, $N_2N$—$CH_2$—$COOH$—the same or different?*

you will find that they are—because glycine does not have an asymmetric carbon atom. There are no enantiomeric forms of glycine, the only amino acid that can claim this distinction.

Now to answer the second question. Enantiomers exhibit the same melting and boiling points, density, solubility, and practically every other physical property. But there is one exception: they affect polarized light differently. In ordinary light, electromagnetic waves (or photons, if you prefer) vibrate in all possible planes simultaneously about a given point. When ordinary light is passed through certain crystals, such as calcite, or a sheet of Polaroid film, the light that emerges consists of waves (or photons) vibrating in only one plane. This is plane-polarized light. When this polarized light is passed through a particular enantiomer, the plane of vibration is rotated one way or the other depending on

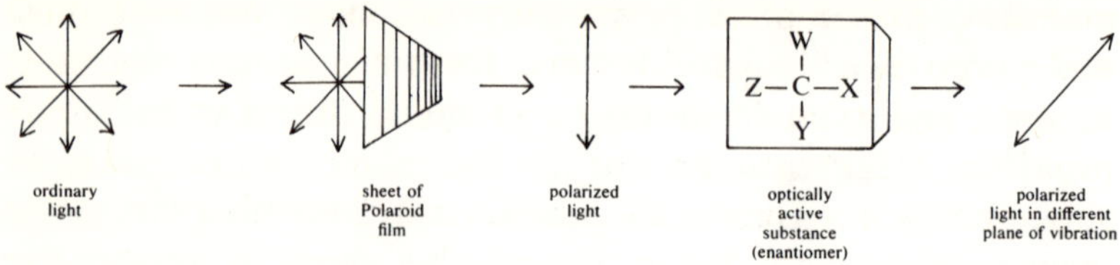

FIG. 69  *The effect of an enantiomer on polarized light.*

which enantiomeric form is used. One enantiomer will rotate the plane of vibration clockwise and is said to be *dextrorotatory* (literally meaning a rotation to the right). The mirror image will rotate the plane of vibration in a counterclockwise direction and is, thus, *levorotatory* (meaning a rotation to the left). In writing the names of enaniantiomers, chemists usually use the symbols d or (+) for the dextrorotatory forms and 1 or (−) for the levorotatory forms. Because of their interaction with polarized light, the term "optical isomers" has also been applied to such molecules. Compounds consisting of one type of enantiomer or optical isomer are said to be *optically active*. Enantiomers are usually found or produced in mixtures containing equal amounts of each form. Such mixtures are called *racemic mixtures*. Inasmuch as each enantiomer rotates light in opposite directions to equal extents, the optical rotation by one enantiomer exactly counterbalances the optical rotation of the other, making the racemic mixture optically inactive. The prefix d, 1, or (±) is, therefore, used to specify a racemic mixture.

Enantiomers also differ from one another chemically in a very subtle way. The enantromeric forms of lactic acid $\left(CH_3-\underset{H}{\overset{OH}{\underset{|}{C}}}-\overset{O}{\overset{\|}{C}}-OH\right)$ ionize to the same degree at the same concentration in water solutions. They are neutralized by sodium hydroxide and can be esterified by ethanol in identical manners. This should be no surprise, since for all intents and purposes each enantiomeric form has exactly the same number and type of atoms in identical arrangements. Approaching reactants encounter the same combination of substituents in either lactic acid form, except that one is the mirror image of the other. There is, in short, no observable difference in chemical behavior when the other reagent is not itself optically active. However, when both reactants have enantiomeric forms, significant differences are observed. It turns out that like forms (two "d"s or two "l"s) tend to react with one another, whereas unlike forms (one "d" with one "l") may not react at all. When d- and l-forms of lactic acid, taken individually in separate containers, react with another optically active reagent, they form activated complexes or transition states which are not mirror images and do not have equal energies. The energy of activation for each reaction is different, with a "d-d" combination of reactants having a lower activation energy than a corresponding "l-d" pair. Consequently, the "d-d" pair should react more rapidly than the "l-d" pair. For this reason, reacting enantiomers of the same type seek out one another, much as one of your hands will search through a pile of gloves until it finds one that fits. These considerations are extremely important in light of the fact that many important biological compounds have enantiomeric forms: amino acids utilized by our bodies are exclusively of the levorotatory form; enzymes and the "substrates" they work on are both optically active and must "fit" together like the pieces of a three-dimensional jigsaw puzzle; only d-glucose can be metabolized by animals and fermented by yeast; l-adrenaline is many times more active than its mirror image.

Enantiomers (optical isomers, mirror images), conformational isomers, and cis-trans isomers are all grouped together under the title of stereoisomerism. You can sometimes observe examples of mirror images at the beach if you are lucky enough to find certain shells. Most snail shells have a right-handed spiral, that is, one which would be traced out by your right hand in a counterclockwise, ascending fashion. Only a few have a left-handed spiral. The two types of spirals are mirror images of each other. On the molecular level, one might assume that the helical proteins and DNA molecules are right-handed, too. The drug LSD also has enantiomeric forms with only the levo form having hallucinogenic powers. An interesting legal battle might ensue if a person were arrested for the sale or use of the non-hallucinogenic d-LSD, since the current law prohibiting the sale of LSD does not specify any particular form.

Near the beginning of this chapter it was stated that eicosane ($C_{20}H_{42}$) has 366,319 isomers. This number refers only to structural isomers, that is, those having a different arrangement of atoms. If we include enantiomeric forms, the number of isomers rises steeply to a total of 3,359,964. The formula $C_{40}H_{82}$ represents $6.25 \times 10^{13}$ structural isomers. Can you imagine how many isomers there must be in all?

# VIII

## Polymers — The Land of the Giants

### The More the Merrier

Many small organic molecules can combine to form extremely large molecules with molecular weights that range into the hundreds of thousands or even millions. This linking together of molecules is called *polymerization*. The very large molecule produced is called a *polymer* (from the Greek: "many parts") and the unit parts are the *monomers*. Polymerization is an important industrial process since it is the source of all plastics, resins, synthetic fibers, lacquers, and paints. Polymerization also occurs in nature whenever animals or plants synthesize proteins, nucleic acids, cellulose, starch, and rubber. The first 150 years of modern chemistry were devoted mainly to the study of low molecular weight compounds. Today, however, much of the research is directed into the area of polymer chemistry.

### Alkenes That Can Add

From Chapter IV, you may recall that alkenes have a tendency to react with other molecules by the process known as addition. Under the right conditions, individual alkenes can also add on to one another. One of the most common polymers is the plastic polyethylene (polyethene), which makes up many household and commercial articles, including food wrapping, garbage bags, wastebaskets, drinking glasses, and bowls, to mention a few. The monomer used to form this polymer is ethylene (ethene), $C_2H_4$, with from 100 to 1,000 of these units being linked to make just one molecule of polyethylene. The molecule of polyethylene may be visualized as a long metal chain or necklace with the individual links as the ethylene monomers. A simplified diagram illustrating the polymerization of ethylene is shown in Figure 70.

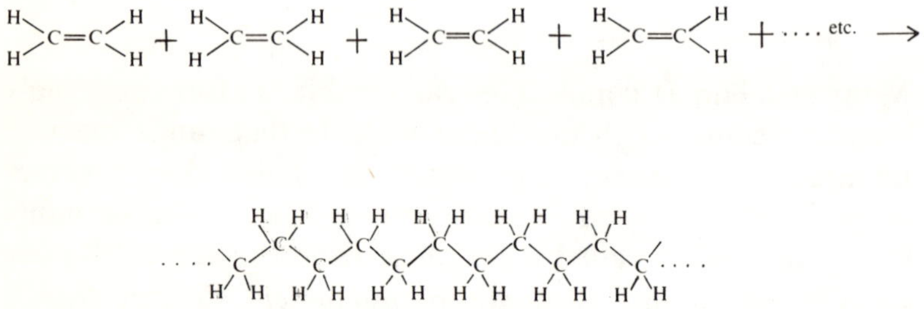

$$\underset{H}{\overset{H}{>}}C=C\underset{H}{\overset{H}{<}} + \underset{H}{\overset{H}{>}}C=C\underset{H}{\overset{H}{<}} + \underset{H}{\overset{H}{>}}C=C\underset{H}{\overset{H}{<}} + \underset{H}{\overset{H}{>}}C=C\underset{H}{\overset{H}{<}} + \cdots \text{etc.} \longrightarrow$$

$$\cdots\underset{H\ H}{\overset{H\ H}{>}}C\underset{H\ H}{\overset{H\ H}{<}}C\underset{H\ H}{\overset{H\ H}{<}}C\underset{H\ H}{\overset{H\ H}{<}}C\underset{H\ H}{\overset{H\ H}{<}}C\underset{H\ H}{\overset{H\ H}{<}}C\underset{H\ H}{\overset{H\ H}{<}}C\cdots$$

FIG. 70   *Polyethylene.*

This diagram is not meant to suggest that the individual monomers all join together at the same time. The actual mechanism is more like a chain reaction. Some type of catalyst is added to initiate the process. It does this by breaking open the double bond on one of the ethylene molecules. Then this ethylene molecule attacks a second ethylene molecule, breaking open its double bond as the two molecules join together to become one. This "doubled-up" molecule repeats the process with a third ethylene molecule, forming a "tripled-up" molecule. The chain reaction is continued in this manner until something happens, such as the linking together of two molecular fragments, to terminate the growth of the polymer. This type of polymerization in which monomers are linked together without any substance being removed is called *addition polymerization*.

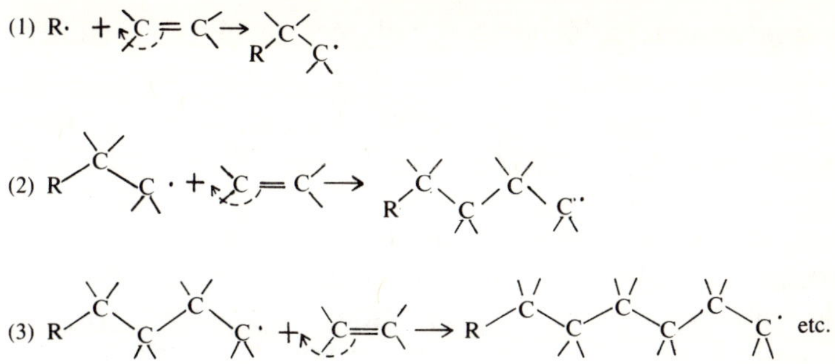

# Polymers—The Land of the Giants

(A) free radical mechanism

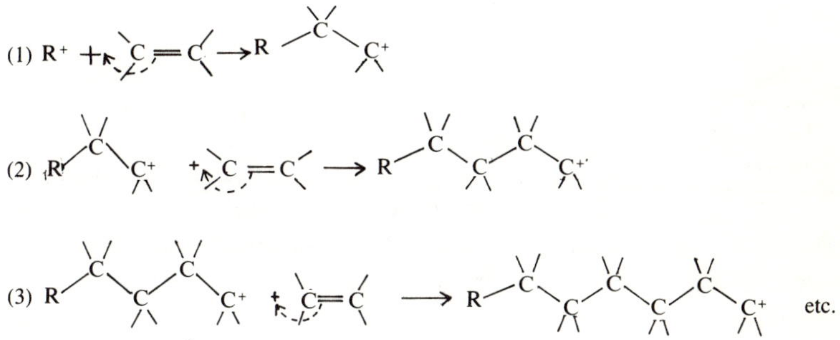

(B) cationic mechanism

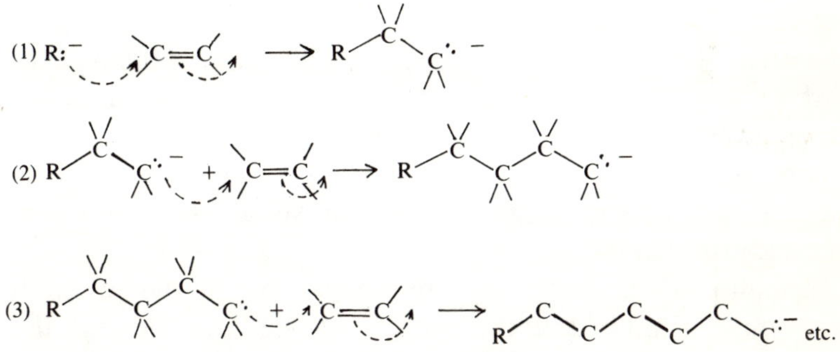

(C) anionic mechanism

FIG. 71 *Three reaction mechanisms for addition polymerization:*
(a) *free radical (a neutral particle with an unpaired electron)*
(b) *cationic (a cation is a positively charged ion)*
(c) *anionic (an anion is a negatively charged ion)*

Any unsaturated molecule can polymerize in this manner. The polymerization of ethylene, like all polymerizations, does not yield one pure, homogeneous product. Many chain reactions may occur simultaneously, and when billions of monomers link together, many strands of polyethylene will be formed, some strands being longer than others. It is impossible, therefore, to write a single chemical equation to adequately represent this reaction. A practical way to write an equation for the polymerization of ethylene is given in Figure 72, where "n" is some large number (from

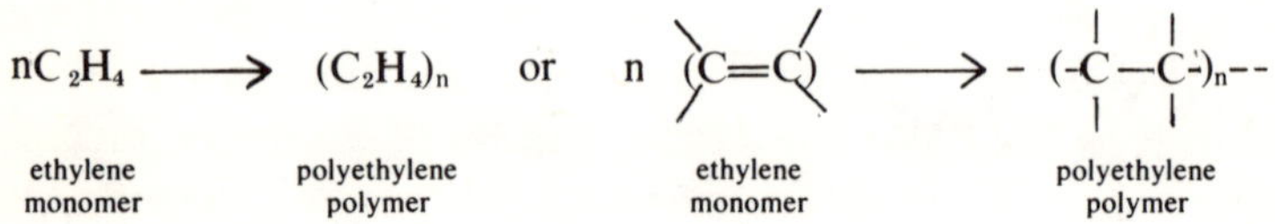

$$nC_2H_4 \longrightarrow (C_2H_4)_n \quad \text{or} \quad n\left(C=C\right) \longrightarrow -(-\overset{|}{\underset{|}{C}}-\overset{|}{\underset{|}{C}}-)_n--$$

ethylene monomer     polyethylene polymer     ethylene monomer     polyethylene polymer

FIG. 72 *Polymerization equation.*

100 to 1,000) of ethylene units. Although polymers are described as "giant" molecules (macromolecules), the largest is actually only a fraction of an inch long when fully stretched out.

Materials made of polyethylene can be made more rigid or more flexible depending on the extent of cross-linking between individual polyethylene chains the manufacturing chemists wish to allow. The cross-links are composed of carbon and hydrogen atoms from the union of side branches of two adjacent polyethylene chains. The length of the chains can be controlled by adjusting conditions such as temperature and pressure and by adjusting the proportions of catalyst to monomer.

A number of important polymers are obtained from monomers similar to ethylene. If we replace one of the hydrogen atoms in ethylene with different substituents, the following polymers can then be formed: polypropylene, polyvinyl chloride or PVC, polystyrene, and polyacrylonitrile (Fig. 73 illustrates the formation of these polymers).

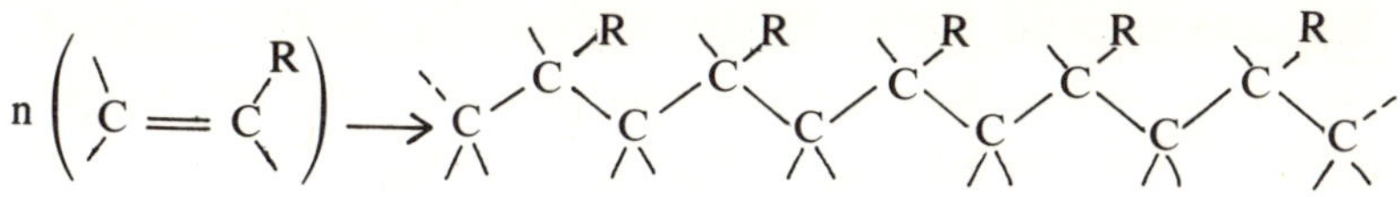

## Polymers—The Land of the Giants

$$n \begin{pmatrix} \diagdown \\ C \\ \diagup \end{pmatrix} = C \begin{pmatrix} R \\ \diagup \\ \diagdown \end{pmatrix} \longrightarrow \cdots C \diagdown_C \diagup^R \diagdown_C \diagup^R \diagdown_C \diagup^R \diagdown_C \diagup^R \diagdown_C \diagup^R \diagdown_C \cdots$$

R = —CH$_3$   polypropylene or polypropene
R = —Cl    polyvinyl chloride or PVC
R = —⬡    polystyrene
R = —C≡N  polyacrylonitrile

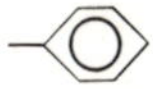

FIG. 73  *Important addition polymers.*

Polypropylene is the material in such items as food bags, indoor-outdoor carpets, and plastic bottles. Polyvinyl chloride is the basis for a large variety of products, including garden hoses, phonograph records, "vinyl" flooring, and disposable plastic bottles and containers. PVC has become a source of environmental concern. When any chlorinated hydrocarbon is incinerated, one product is hydrogen chloride. In the atmosphere, the hydrogen chloride unites with water to form hydrochloric acid. This is harmful to plants and animals and speeds up the corrosion of items containing metal parts. Today, many scientists fear that the manufacture of items made of PVC may cause an industrial cancer epidemic among the workers concerned. Legislators may soon require these workers to wear protective devices, including gas masks. The increased costs of manufacturing PVC would then, naturally, be passed on to you and me, the consumers.

Polystyrene is molded into many different shapes and is found in a variety of household items. When air is mixed into the liquid polystyrene before it cools, a foam called styrofoam is produced. This material has excellent insulating properties and is fashioned into cups and space fillers in packaging. Polyacrylonitrile is more familiarly known as Orlon when it is spun into fibers.

Replacing two hydrogen atoms, both bonded to the same

carbon, gives rise to a number of other familiar polymers, including polyisobutene (the adhesive in "Band-Aids"), Saran (plastic wrap), and Lucite or Plexiglas (used as a glass substitute in windows and bottles and also in "latex" paints). The monomers for these polymers are shown in Figure 74. The molecules of natural rubber (polyisoprene) are now known to have the very regular array shown in Figure 75. Actually, these chains are not straight as shown here, but coiled, twisted, and intertwined with each other, resembling, to some extent, spaghetti in a bowl. When the rubber is stretched, the coils are extended and straightened. This action creates a higher energy, less stable state. When the rubber is allowed to relax, the coiled condition is resumed. The process of vulcanization was discovered by Charles Goodyear when he accidentally spilled some sulfur into a vat of raw rubber. The addition of the sulfur was found to enhance the rigidity, resistance to heat, and strength of

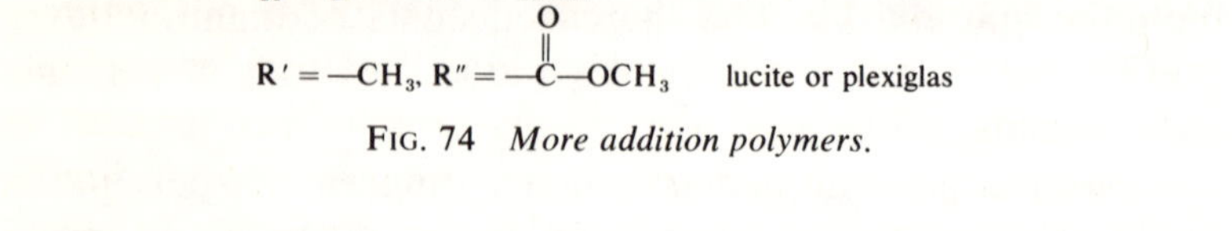

$R' = R'' = -CH_3$    polyisobutylene or polyisobutene
$R' = R'' = -Cl$    saran

$R' = -CH_3, R'' = -\overset{O}{\underset{\|}{C}}-OCH_3$    lucite or plexiglas

FIG. 74  *More addition polymers.*

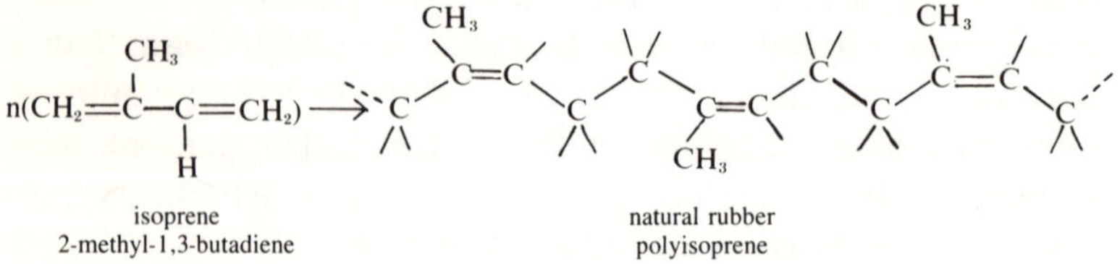

$n(CH_2=\underset{\underset{H}{|}}{\overset{\overset{CH_3}{|}}{C}}-C=CH_2) \rightarrow$

isoprene
2-methyl-1,3-butadiene

natural rubber
polyisoprene

FIG. 75  *Structure of natural rubber.*

the rubber. These properties are explained by the fact that the sulfur atoms form cross-links between adjacent chains of rubber molecules (Figure 76). The rigidity of the rubber increases as the sulfur content increases. Hard rubber containing up to 35 percent sulfur is used to manufacture automobile battery cases.

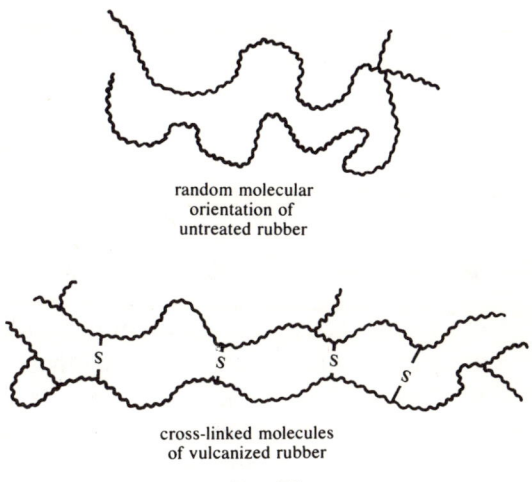

random molecular orientation of untreated rubber

cross-linked molecules of vulcanized rubber

FIG. 76

It is interesting to note that the isoprene unit is one of nature's favorite building blocks. It is found not only in rubber, but in many other compounds obtained from plants and animals. Terpenes, which are compounds present in the essential oils of various plants, contain isoprene units, as do vitamin A and retinal (notice the similarity in structure). Both vitamin A (see Chapter V) and retinal play a vital role in the chemistry of vision. Although not as evident, other molecules such as cholesterol are also derived from isoprene units.

Teflon is a plastic frequently applied as a coating on such items as machine tools and frying pans. Teflon is noted for its chemical inertness and its low-friction surface. Its monomer is tetrafluoroethylene (tetrafluoroethene), which resembles etheylene made up of four fluorine atoms instead of four hydrogen atoms. It polymerizes in the same manner as ethylene (Figure 77).

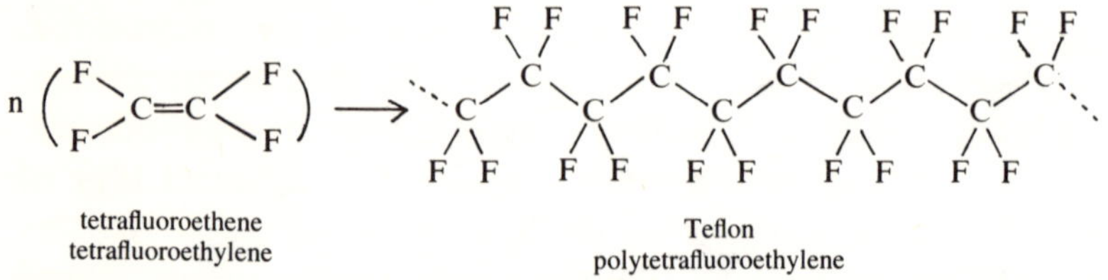

$$n \left( \underset{F}{\overset{F}{>}} C = C \underset{F}{\overset{F}{<}} \right) \longrightarrow \cdots \underset{F\ F}{\overset{F\ F}{|}} C - \underset{F\ F}{\overset{F\ F}{|}} C - \underset{F\ F}{\overset{F\ F}{|}} C - \underset{F\ F}{\overset{F\ F}{|}} C - \underset{F\ F}{\overset{F\ F}{|}} C - \underset{F\ F}{\overset{F\ F}{|}} C \cdots$$

tetrafluoroethene
tetrafluoroethylene

Teflon
polytetrafluoroethylene

FIG. 77 *Polymerization of Teflon.*

## Copolymers: Where Molecules Live Together

The polymers mentioned so far are all of the addition type. Since only one kind of monomer was used, these polymers may be called *homopolymers* (*homo* means "the same"). Those which are synthesized by the linking together of repeating units consisting of different kinds of monomers are called *copolymers*. The process of copolymerization may be represented in a simplified fashion, as shown in Figure 78.

The distribution of each kind of monomer, however, can range from complete randomness to strict alternation along the chain. A particularly important copolymer is the one formed from the monomers 1,3-butadiene (containing two

(a)  A+A+A+A+A+A····⟶ A—A—A—A—A—A····

(b)  A+B+A+B+A+B----⟶ A—B—A—B—A—B - - - -

FIG. 78 *Types of polymers:*
(a) *a homopolymer (one kind of monomer)*
(b) *a copolymer (more than one kind of monomer)*

double bonds) and styrene. This copolymer consists of approximately three parts of 1,3-butadiene to one part of styrene and is known as **SBR**. During World War II, SBR was the most important of the synthetic rubbers developed to replace natural rubber, which was unavailable when the United States was cut off from its Far Eastern sources. Both the 1,3-butadiene and the styrene monomers are unsaturated, and the synthesis of SBR was one of addition polymerization.

Nylon was one of the first synthetic fibers to be produced. It, too, is a copolymer, but it is made by a different type of polymerization called *condensation polymerization*. In this type of polymerization, double bonds are not required and some small molecules (usually water or ammonia) are eliminated from the reacting monomers as they "condense" to form the polymer. A condensation polymer can be made only from monomers that are, at least difunctional (containing two functional groups) so that each kind of monomer can react with the other kind at both ends of the molecule. Only in this way can a giant molecule be developed. To illustrate this point, consider the reaction of a monohydric alcohol with a monocarboxylic acid, as shown in Figure 79.

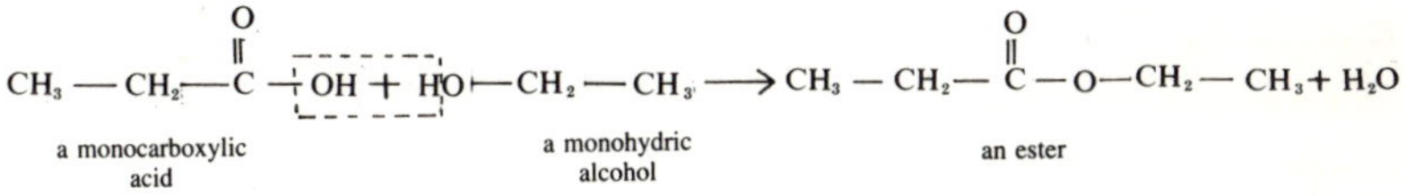

$$CH_3-CH_2-\overset{\overset{O}{\|}}{C}-OH + HO-CH_2-CH_3 \longrightarrow CH_3-CH_2-\overset{\overset{O}{\|}}{C}-O-CH_2-CH_3 + H_2O$$

a monocarboxylic acid      a monohydric alcohol      an ester

FIG. 79

This condensation does not result in a polymer because once the functional groups have reacted, there are no other functional groups to sustain the reaction. Now, if a dihydric alcohol or "diol" reacts with a monocarboxylic acid, as shown in Figure 80, the product will still have another hydroxyl group and can, therefore, react with another molecule of acid. This gives us a "polymer" consisting of only three

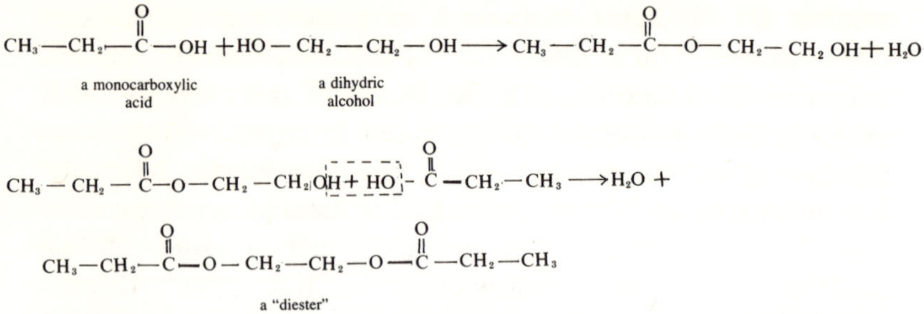

104 THE STUDENT CHEMIST EXPLORES ORGANIC COMPOUNDS

$$CH_3-CH_2-\overset{\overset{O}{\|}}{C}-OH + HO-CH_2-CH_2-OH \longrightarrow CH_3-CH_2-\overset{\overset{O}{\|}}{C}-O-CH_2-CH_2OH + H_2O$$

a monocarboxylic acid    a dihydric alcohol

$$CH_3-CH_2-\overset{\overset{O}{\|}}{C}-O-CH_2-CH_2|\overline{OH + HO}|-\overset{\overset{O}{\|}}{C}-CH_2-CH_3 \longrightarrow H_2O +$$

$$CH_3-CH_2-\overset{\overset{O}{\|}}{C}-O-CH_2-CH_2-O-\overset{\overset{O}{\|}}{C}-CH_2-CH_3$$

a "diester"

FIG. 80

units (called a trimer). By this time it should be apparent that in order to get a long chain consisting of many units, each monomer must have two functional groups. If we take a dihydric alcohol, such as ethylene glycol (ethanediol), and a dicarboxylic acid, such as terephthalic acid, the reaction shown in Figure 81 occurs.

Since an alcohol and acid combine to form an ester, the condensation of many alcohol monomers and many acid monomers produces a "polyester." Many water molecules are eliminated in the process, one water molecule being released for each "ester linkage" formed. When processed into fibers, the polyester illustrated in Figure 81 is commonly known as Dacron.

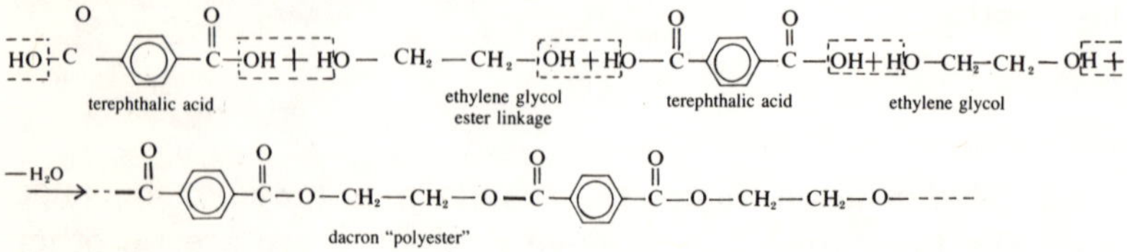

$$\overline{HO}-C-\underset{\text{terephthalic acid}}{\bigcirc}-\overset{\overset{O}{\|}}{C}-\overline{OH + HO}-\underset{\text{ethylene glycol}}{CH_2-CH_2}-\overline{OH + HO}-\overset{\overset{O}{\|}}{C}-\underset{\text{terephthalic acid}}{\bigcirc}-\overset{\overset{O}{\|}}{C}-\overline{OH + HO}-\underset{\text{ethylene glycol}}{CH_2-CH_2}-\overline{OH} +$$

ester linkage

$$\xrightarrow{-H_2O} \cdots -\overset{\overset{O}{\|}}{C}-\bigcirc-\overset{\overset{O}{\|}}{C}-O-CH_2-CH_2-O-\overset{\overset{O}{\|}}{C}-\bigcirc-\overset{\overset{O}{\|}}{C}-O-CH_2-CH_2-O-\cdots$$

dacron "polyester"

FIG. 81

A type of nylon called nylon 66 is made from the monomers adipic acid, a dicarboxylic acid, and hexamethylene diamine or 1,6-hexanediamine (containing two amine groups). The

## Polymers — The Land of the Giants

number 66 indicates that each monomer consists of six carbon atoms. The condensation of these monomers releases molecules of ammonia, and the union of the carboxyl and amino groups produces the "amide" linkage. Nylon is an example of a "polyamide." The molecular weight of nylon may be as high as 25,000. Individual strands of nylon are able

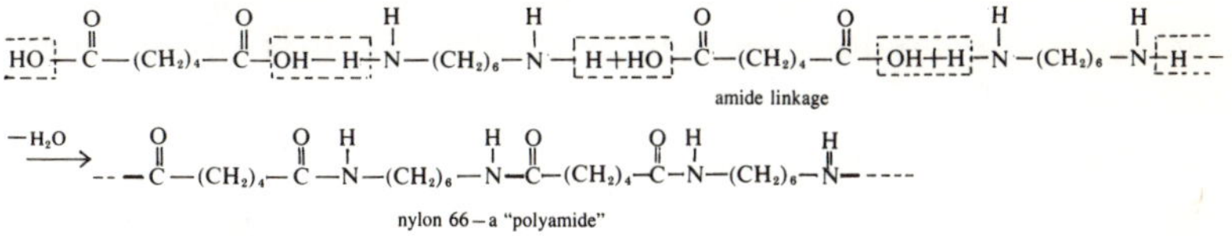

$$\text{HO--C--(CH}_2)_4\text{--C--OH} + \text{H--N--(CH}_2)_6\text{--N--H} + \text{HO--C--(CH}_2)_4\text{--C--OH} + \text{H--N--(CH}_2)_6\text{--N--H}$$

amide linkage

$$\xrightarrow{-H_2O} \text{--C--(CH}_2)_4\text{--C--N--(CH}_2)_6\text{--N--C--(CH}_2)_4\text{--C--N--(CH}_2)_6\text{--N--}$$

nylon 66 — a "polyamide"

FIG. 82

to stick together because of the very strong attraction between the hydrogen of an "-NH-" group in one strand and an oxygen from a carbonyl group in an adjacent strand. These strong bonds, called hydrogen bonds, give the nylon fabrics and cords great strength and durability.

Phenol, $C_6H_5OH$, and methanol (formaldehyde) are the monomers used to synthesize Bakelite, one of the first plastics made and still one of the most widely used. Uses of Bakelite range from telephones to insulators to pipe stems to radios. Bakelite's high rigidity and resistance to heat are due to the three-dimensional, highly cross-linked nature of the plastic.

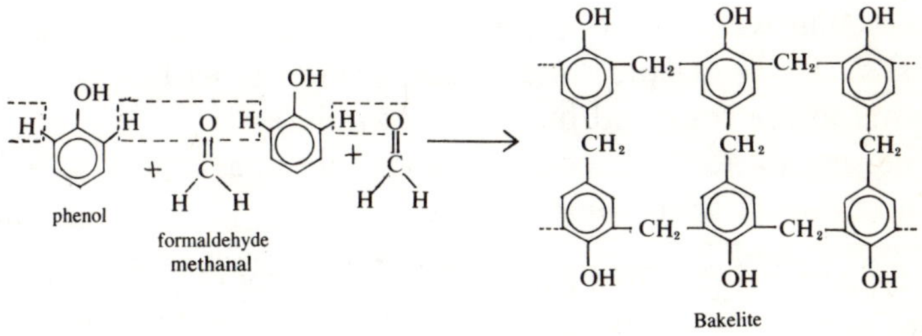

FIG. 83

## Building a Building Block

Proteins are often referred to as the "building blocks of nature." Every part of the human body contains proteins, some parts containing more than others. Nucleic acids control the behavior of our cells by means of the particular proteins they allow to be formed. This is true not only for humans, but for all plants and animals. Every living organism, including the smallest viruses, contains protein. Furthermore, the kinds of proteins vary from one type of living organism to another. The proteins found in the blood of an octopus, for example, are different from those in the fingernail of a monkey. Even within a given type of organism, proteins can vary from one species to another. Consequently, the number of proteins that occur in nature is virtually infinite. The enormous number of proteins enables us to understand why there is such an infinite variety of living organisms in nature. But how is it possible for so many proteins to come to be? To answer this question we must first investigate how proteins are made.

Proteins are polymers, containing the same type of linkage — the amide linkage — found in nylon. The monomers are the amino acids, of which there are about twenty, varying in size from a molecular weight of 90 to one of about 250. An average protein contains approximately 500 amino acids, but there are many proteins containing far more than this number and others with far fewer. The average protein has a molecular weight of 60,000. Many go much higher. In some viruses, for example, there are proteins with molecular weights in the tens and hundreds of millions!

The amino acid molecules are difunctional, each containing a carboxyl group and an amino group. Amino acids may be represented by the general formula shown in Figure 84. The different amino acids are obtained by substituting different organic groups for "R." For example, if $R = H$, the amino acid is glycine; if $R=CH_3$, it is alanine; if $R=(CH_3)_2 CHCH_2$, it is leucine; and so on. The monomers condense by eliminating water molecules and then by joining together at the sites vacated by the H and OH particles, as shown in Figure 85.

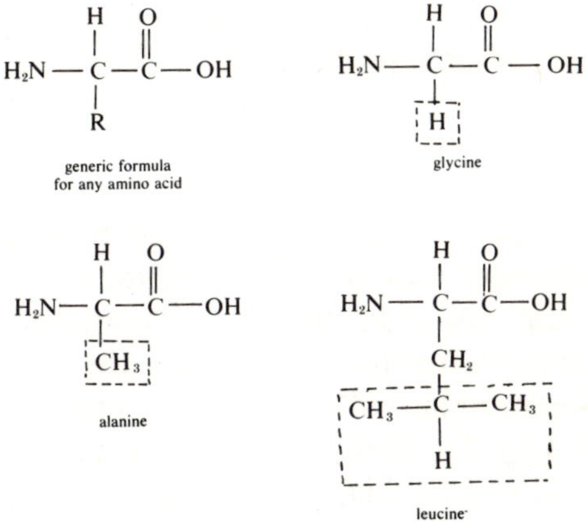

FIG. 84  *Amino acid formulas.*

In order to answer our original question, we have to determine how many different combinations or sequences of amino acids are possible in building a typical protein molecule. Suppose we start with a hypothetical "protein" containing only two amino acids,[1] A and B. The two sequences that we can write, $A - B$ and $B - A$, are actually only one molecule since molecules can be turned around or viewed from different sides. With three amino acids (A, B, and C), we can get three different molecules (ABC or CBA, ACB or BCA, and BAC or CAB). With four amino acids, we can get twelve different molecules. With the aid of mathematics, the range

---
[1] Such a molecule is far too small to be regarded as a protein.

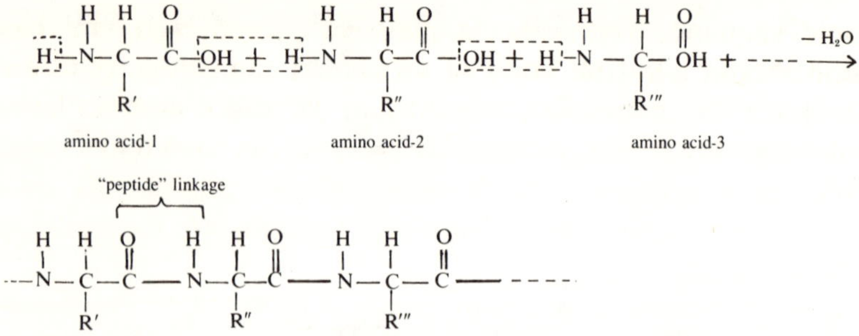

## 108 THE STUDENT CHEMIST EXPLORES ORGANIC COMPOUNDS

$$\underset{\text{amino acid-1}}{H_2N-\underset{\underset{R'}{|}}{\overset{\overset{H}{|}}{C}}-\overset{O}{\overset{\|}{C}}-OH} + \underset{\text{amino acid-2}}{H-N-\underset{\underset{R''}{|}}{\overset{\overset{H}{|}}{C}}-\overset{O}{\overset{\|}{C}}-OH} + \underset{\text{amino acid-3}}{H-N-\underset{\underset{R'''}{|}}{\overset{\overset{H}{|}}{C}}-\overset{O}{\overset{\|}{C}}-OH} + \cdots \xrightarrow{-H_2O}$$

$$\underset{}{\cdots-N-\underset{\underset{R'}{|}}{\overset{\overset{H}{|}}{C}}-\overset{O}{\overset{\|}{C}}}\overbrace{-N-\underset{\underset{R''}{|}}{\overset{\overset{H}{|}}{C}}-\overset{O}{\overset{\|}{C}}}^{\text{"peptide" linkage}}-N-\underset{\underset{R'''}{|}}{\overset{\overset{H}{|}}{C}}-\overset{O}{\overset{\|}{C}}-\cdots$$

FIG. 85  *Protein synthesis — "R" represents an organic group which varies with each different amino acid.*

of possible molecules can be calculated by multiplying the number of different amino acids, labelled "n," by (n − 1), and then multiplying again by (n − 2), and continuing the process of multiplying until we reach the number 1 as a multiplier. The product [of n times (n − 1) times (n − 2 . . .] is then divided by two (because each molecule is the same in reverse). To illustrate this, if we had six amino acids, the number of possible molecules obtainable is (6 × 5 × 4 × 3 × 2 × 1) ÷ 2, or 360 possible structures. This method assuredly beats trial and error! By the time you get to ten amino acids, there are nearly 2,000,000 possibilities. If we use twenty amino acids, almost $1.25 \times 10^{18}$ arrangements are possible even though we have used each amino acid monomer only once. In our typical protein with 500 amino acid monomers, only about twenty different amino acids are present. It follows, therefore, that the same amino acids must be used more than once. The number of possible proteins now will be far, far greater than $1.25 \times 10^{18}$. The number of variations may even exceed the total number of atoms in the universe. This may not be so surprising when you think of how many words have been and could be made from the twenty-six letters of the alphabet. Remember that each different amino acid sequence represents a different protein and you can understand why the body of an animal can design different proteins to accomplish a multitude of tasks without ever running out of new varieties. Now you can also understand why each species of each organism can produce its own characteristic proteins. No wonder proteins are the building blocks of life.

Much current scientific research is devoted to the determination of the sequences of amino acids in various proteins. Once the sequence is established, the scientists then hope to build the protein synthetically. The complete amino acid sequence of insulin, the hormone used to treat diabetes, was elucidated in 1954. It has a molecular weight of 5,734 and contains forty-eight amino acid units of sixteen different amino acids. The most important proteins are probably the *globular proteins,* in which the chains are not merely straight lines, but exist in complex loops and twists held together in this condition by hydrogen bonds in much the same manner as in nylon. Some proteins contain sulfur, which, as in vulcanized rubber, provides for the cross-linking of individual chains.

## *Polymers Made of Sugar*

Starch and cellulose are natural polymers consisting of the same monomer: glucose, $C_6H_{12}O_6$ (actually d-glucose). There are approximately 1,500 glucose units in a cellulose molecule, whereas a starch molecule has only about 220 glucose units. They differ also in the way the individual glucose units are joined together and in the degree of cross-linking between chains.

Glucose, itself, is classified as a simple sugar or *monosaccharide,* which means that it cannot be made to yield any simpler carbohydrate. Starch and cellulose are *polysaccharides,* which, upon reaction with water, are broken down into their component glucose units. The molecule of glucose has been found to have a cyclic structure that is known to exist in two possible forms, as shown in Figure 86. The only

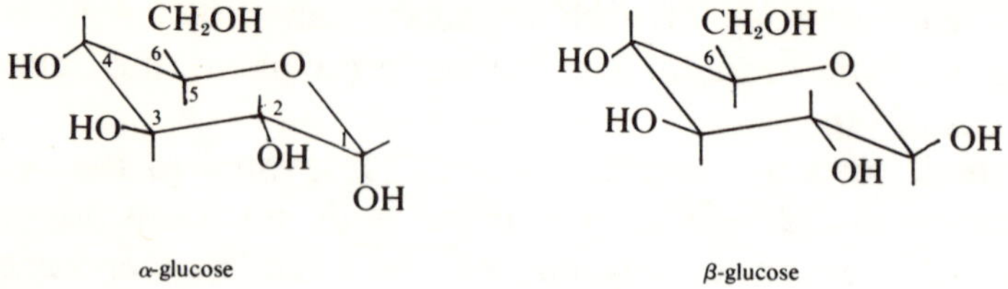

α-glucose     β-glucose

FIG. 86

difference between these forms is that in α-glucose, the OH group on carbon #1 is axial, whereas in β-glucose the OH group is equatorial. It is the α-glucose molecules that are joined together in starch and the β-glucose molecules that are joined together in cellulose.

A number of other molecules are isomers of glucose. One

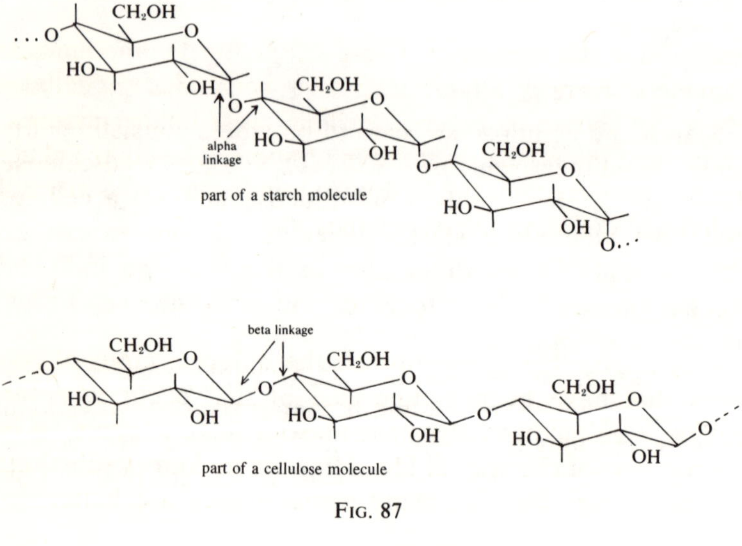

part of a starch molecule

part of a cellulose molecule

FIG. 87

of these, fructose, combines with glucose to form sucrose, which is ordinary table sugar.

Cotton fabric, made of cellulose chains, has been given a "permanent press" feature by the addition of compounds known as dimethylols (meaning two —$CH_2$ OH groups at

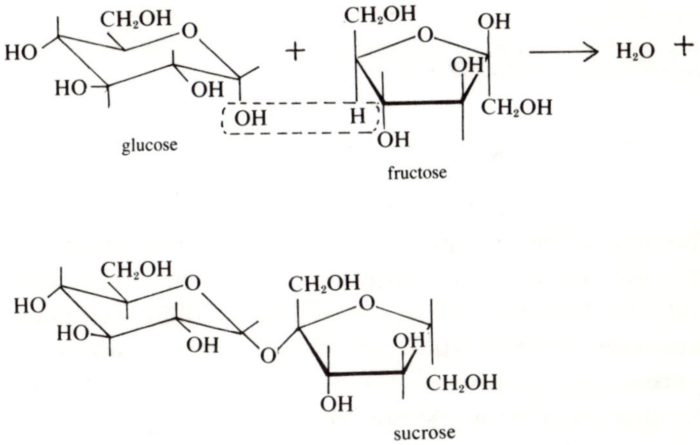

FIG. 88  *Sucrose.*

each end of the molecule). These difunctional compounds are capable of forming ether linkages between adjacent cellulose fibers, as shown in Figure 89. This cross-linking assures a permanent interchain orientation of fibers. The ether linkages are far stronger than the hydrogen bonds that ordinarily hold individual cellulose fibers together.

## Future Shock

It is impossible to mention all the different kinds of polymers that have been synthesized or that occur in nature. Suffice it to acknowledge that plastics have become an important part of our way of life. They are no longer substitutes for metals or glass. In many respects they are better than

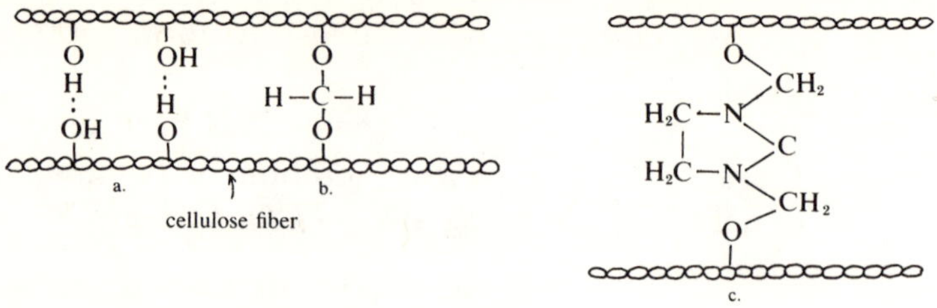

FIG. 89  Cross-linking in cellulose fibers.
(a) *natural cross-linking — hydrogen bonds*
(b) *and* (c) *cross-linking initiated by dimethylolurea and dimethylol ethyleneurea, respectively*

these materials. Soluble plastic bottles are being developed that, like ice cream cones once they have served their purpose, disappear. These plastics, of course, must be biodegradable in order not to create additional environmental headaches. Human bodies, like our automobiles, will contain increasingly more plastic parts. Heart valves constructed of polypropylene can flex 40,000,000 times a year without noticeable wear. A water-absorbing plastic, which will allow the passage of tears from the cornea of the eye, has been used to make contact lenses. Human eye lenses, too, soon may be replaced by special inert plastic ones. A polymer called polyethylene oxide (the monomer is ethylene oxide) added to water in low concentration causes the water to act as though it had been lubricated. This slippery water, called "polywater," can be used to lengthen fire hoses without loss of nozzle pressure. The longer hoses and the easier flowing water could make fire-fighting more effective.

The boom in the plastics industry has created many environmental problems. During plastics manufacture, significant amounts of gaseous and liquid monomers escape into the atmosphere. The solvents used in paints evaporate and further pollute the air as the paint dries. Waste products from factories are expelled into our waters. Plastics with low biodegradability can remain undecomposed for hundreds of years. Incineration of plastics, as in the case of PVC, can create health problems. New and more potent drugs and pesticides are constantly being synthesized. Each year there are significant breakthroughs in the area of biochemical genetics. Having already synthesized a number of proteins, perhaps scientists soon will be able to synthesize nucleic acids and determine at will what characteristics a human offspring should and should not have.

These are but a few of the problems that organic chemists and other scientists will have to face in the future. Although you are a student chemist now, you may someday be in a position to play a vital role in the solution of these difficult problems. Maybe you can do something to make our world a better place in which to live—with clear skies above and clean water below—a world in which all of us can enjoy our "seasons in the sun."

Accept this as a challenge. If you don't, who will?

# For Further Reading

Aurand, L. W., and Woods, A. E. *Food Chemistry.* Westport, Conn.: AVI Publishing Company, 1973.

Benfey, O. Theodor. *From Vital Force to Structural Formulas.* Boston, Mass.: Houghton Mifflin Company, 1964.

———. *The Names and Structures of Organic Compounds.* New York, N.Y.: John Wiley & Sons, Inc., 1966.

Ferguson, Lloyd N. *Organic Chemistry.* Boston, Mass.: Willard-Grant Press, Inc., 1972.

———. *Organic Molecular Structure.* Boston, Mass.: Willard-Grant Press, Inc., 1974.

Hart, Harold, and Schultz, Robert D. *Organic Chemistry; A Short Course.* 4th Ed. Boston, Mass.: Houghton Mifflin Company, 1972.

Hill, John W. *Chemistry for Changing Times.* Minneapolis, Minn.: Burgess Publishing Company, 1972.

Holum, John R. *Elements of General and Biological Chemistry.* 3rd Ed. New York, N.Y.: John Wiley & Sons, Inc., 1972.

Jacobson, Martin, and Beroza, Morton. "Insect Attractants," *Scientific American,* August 1964, Vol. 211, No. 2.

Jarvis, Bruce, and Mazzocchi, Paul. *Form and Function— An Organic Chemistry Module.* New York, N.Y.: Harper & Row Publishers, 1973.

Kermode, G. O. "Food Additives," *Scientific American,* March 1972, Vol. 226, No. 3.

Lykken, Louis. "Chemical Control of Pests," *Chemistry,* July–August 1971, Vol. 44, No. 7.

Mills, G. Alex. "Ubiquitous Hydrocarbons, Part I," *Chemistry,* February 1971, Vol. 44, No. 2.

———. "Ubiquitous Hydrocarbons, Part II," *Chemistry,* March 1971, Vol. 44, No. 3.

Morrison, Robert T., and Boyd, Robert N. *Organic Chemistry.* 3rd Ed. Boston, Mass.: Allyn & Bacon, Inc., 1973.

Naves, Renoe G., and Strickland, Barbara. "Barbiturates," *Chemistry,* March 1974, Vol. 47, No. 3.

Ramsey, O. Bertrand. "Molecules in Three Dimensions, Part I," *Chemistry,* January 1974, Vol. 47, No. 1.

Rouvray, Dennis. "The Mathematical Theory of Isomerism," *Chemistry,* February 1972, Vol. 45, No. 2.

Schaumberg, Gene D. *Concerning Chemistry.* New York, N.Y.: John Wiley & Sons, Inc., 1974.

Schneider, Dietrich. "The Sex Attractant Receptor of Moths," *Scientific American,* July 1974, Vol. 231, No. 1.

Schubert, Leo, and Veguilla-Berdecia, L. A. *Chemistry and Society.* Boston, Mass.: Allyn & Bacon, Inc., 1972.

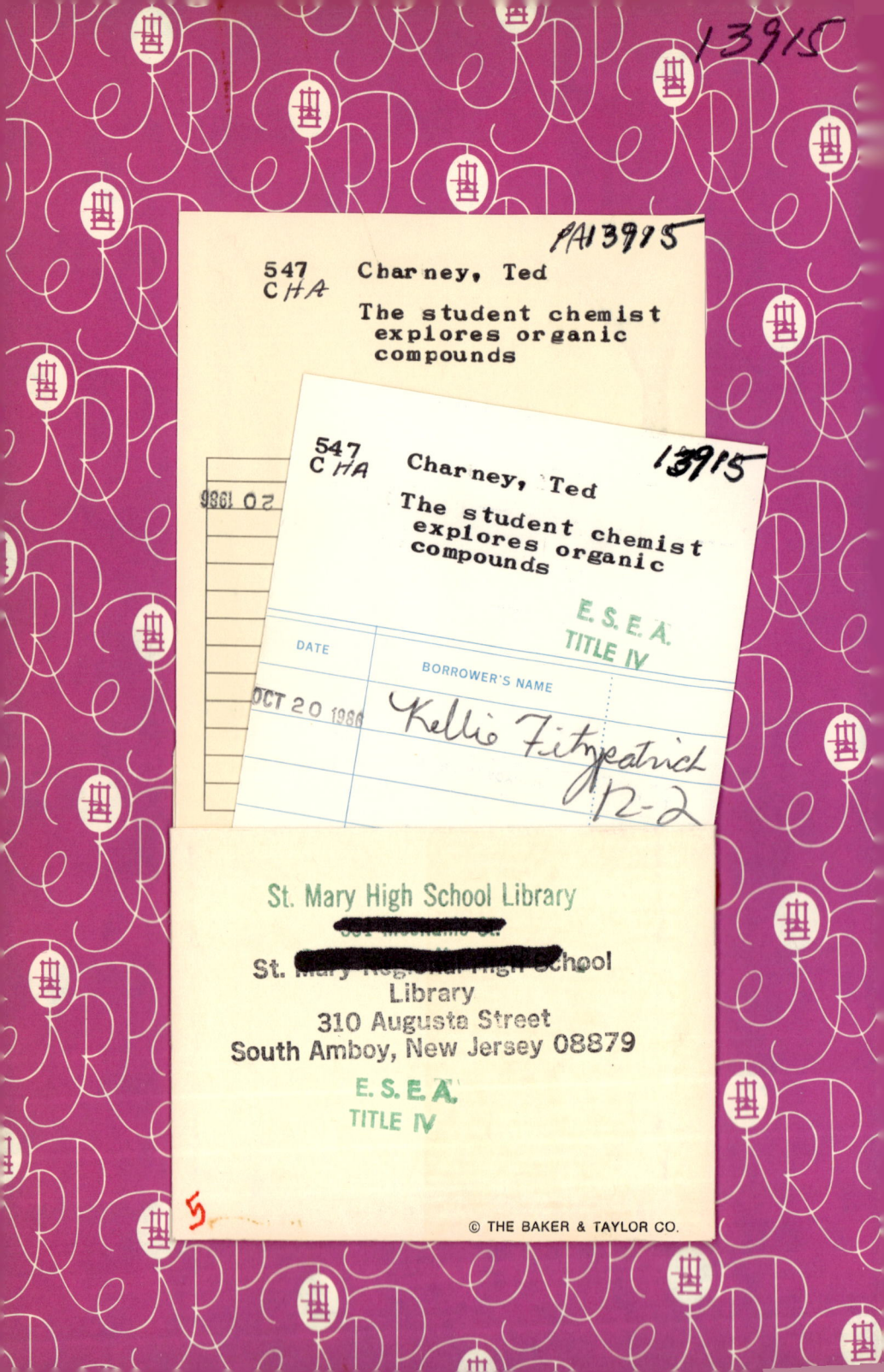